THE AGE OF UNCERTAINTY

Tobias Hürter, born in 1972, studied mathematics and philosophy in Munich and Berkeley, and has been writing about science, technology, and philosophy for magazines and newspapers since 2000. He worked as an editor at the *MIT Technology Review* and was deputy editor at *Hohe Luft*, a philosophy magazine that he co-founded. He is now a permanent freelance editor at *DIE ZEIT Magazin Wissen*. Hürter is the author of several nonfiction books.

David Shaw works as a journalist for Germany's international broadcaster, Deutsche Welle, as well as translating from several languages, including German, Dutch, Russian, and French. He lives in Berlin.

AF583816

THE AGE OF UNCERTAINTY

how the greatest minds in physics changed the way we see the world

Tobias Hürter

Translated by David Shaw

SCRIBE
Melbourne • London

Scribe Publications
18-20 Edward St, Brunswick, Victoria 3056, Australia
2 John St, Clerkenwell, London, WC1N 2ES, United Kingdom

First published in German as *Das Zeitalter der Unschärfe* by Klett-Cotta in 2021
This edition published by arrangement with Michael Gaeb Literary Agency, Berlin
Published by Scribe in 2022

Text copyright © 2021 by Klett-Cotta – J.G. Cotta'sche Buchhandlung Nachfolger GmbH, Stuttgart/Germany
Translation copyright © David Shaw 2022

All rights reserved. Without limiting the rights under copyright reserved above, no part of this publication may be reproduced, stored in or introduced into a retrieval system, or transmitted, in any form or by any means (electronic, mechanical, photocopying, recording or otherwise) without the prior written permission of the publishers of this book.

Typeset in Adobe Caslon Pro by the publishers

Printed and bound in Australia by Griffin Press

Scribe is committed to the sustainable use of natural resources and the use of paper products made responsibly from those resources.

Scribe acknowledges Australia's First Nations peoples as the traditional owners and custodians of this country. We recognise that sovereignty was never ceded, and we pay our respects to their elders, past and present.

978 1 922585 50 9 (Australian edition)
978 1 914484 42 1 (UK edition)
978 1 922586 69 8 (ebook)

Catalogue records for this book are available from the National Library of Australia and the British Library.

scribepublications.com.au
scribepublications.co.uk

To Herbert Schmidt

Contents

Prologue

Imagine one day you found out that the world you live in works completely differently from the way you thought. The buildings, streets, trees, clouds are nothing but pieces of theatre scenery, shifted by forces you never dreamed existed.

That's exactly what happened to physicists a century ago. They were forced to accept the fact that, behind the concepts and theories through which they saw the world, there lay a deeper reality. And it was one that seemed so strange to them that they began to argue about whether it still made sense to speak of any kind of 'reality' at all.

How those physicists came to such a realisation, and how they subsequently wrestled with it, is the story of this book. By the end of the story, the world will have become a different place, reinterpreted and fundamentally altered by those physical scientists.

PARIS, 1903

Cracks begin to appear

Paris, 1903. A summer evening in June. In a garden on the Boulevard Kellermann in the thirteenth arrondissement, light streams through windows, illuminating a lawn. A door opens, and the sound of merry voices emerges, followed by a small group of party guests. They spill out onto the gravel pathways, a woman wearing a black dress in their midst: the thirty-nine-year-old physicist Marie Curie. Her often-tense expression is now relaxed and happy. She's hosting this party to celebrate gaining her doctorate.

Marie is at the pinnacle of her career. She's the first woman in France to be awarded a doctorate in natural science, which she achieved with the distinction '*très honorable*'. And she's the first woman ever to be nominated for the Nobel Prize.

Marie's husband, Pierre, stands at her side, beaming with pride. Marie is surrounded by her elder sister, Bronya, her doctoral supervisor, Gabriel Lippmann, fellow physicists Jean Perrin and Paul Langevin, and several young women, who are her students. The New Zealand–born physicist Ernest Rutherford is also among the guests at the party. He and his wife are currently on their honeymoon — finally, as they've already been married for three years. Ernest Rutherford and Marie Curie are rivals in studying the structure of the atom, about which they vehemently disagree. But those arguments have been put aside for the evening. This is a time to celebrate.

The path that eventually leads Marie to this evening of celebration begins far from the French capital, in 1860s Warsaw. Poland is divided among the great powers of Prussia, Russia, and Austria, and Warsaw is under the yoke of the Russian czars. Poles are forbidden from using the name 'Poland' to refer to their homeland. This is where Maria Skłodowska is born on 7 November 1867. She's the youngest of five children, and both her parents are teachers. The family is opposed to the Russian occupation. Their father does his best to raise his daughters as independent thinkers. When Manya, as she's known affectionately at home, is four years old, her mother withdraws from the family to avoid infecting them with tuberculosis. Contact is kept to a minimum, and she eventually dies after a long struggle with the then-incurable disease.

It will be ten years before Manya recovers her lust for life. At first, she finds escape in learning, burying herself in her books, and, thanks to her relentless hard work, she graduates from the Imperial Secondary School at the top of her class. When she's fifteen years old, all that self-imposed pressure becomes too much, and she suffers a nervous breakdown. Now a single parent, her father sends her to the country to recover. There, she manages to set aside her books, and discovers music and parties, dancing and flirting through the night. She begins studying at an underground Polish university that accepts women, where she easily outperforms all her fellow students. When Bronya, her elder sister by two years, moves to Paris to study medicine, Manya takes a position as a governess to the children of a sugar-beet tycoon to help pay for Bronya's studies. There, she falls in love with the adult son of the family, a twenty-three-year-old mathematics student called Casimir. The young man's father is horrified when he hears of their affair. At first, Casimir tentatively resists his father, but after several years of vacillating, he eventually gives in, leaving Manya bereft and alone, with a deeply wounded heart and a passionate anger directed at all men. 'If they don't want to marry impecunious young girls, let them go to the devil!'

In 1891, Manya joins her sister in Paris. Bronya is now married — to a man called Casimir, as fate would have it. Both Bronya and her husband are doctors, and both are fired with communist ideals. They practise medicine from their apartment, treating poor and needy patients for free. The commotion is too much for her, and Manya, or

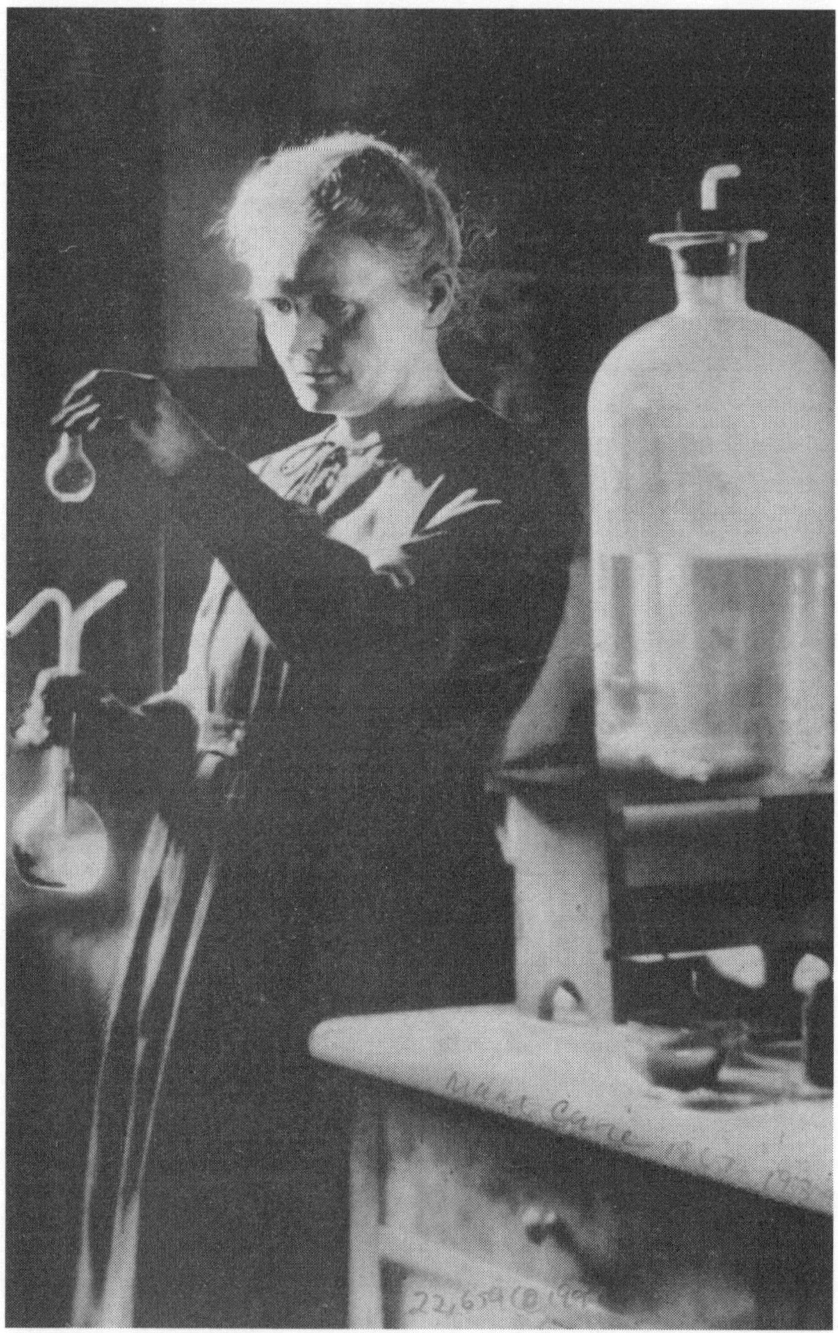

Marie Curie received two Nobel Prizes. She was awarded the Nobel Prize for Physics in 1903, and the Prize for Chemistry in 1911. This picture shows her in her Paris laboratory in 1917.

Marie, as she now calls herself, moves into a rented garret, where she literally buries herself under every item of clothing she owns to keep warm through the cold winter nights. Needing to save money, she rarely hauls a bucket of coal up to her room, and lives on tea, fruit, dry bread, and chocolate—but none of it matters! She's free now. Women are anything but equal citizens in turn-of-the-century Paris, where the word *étudiante* can refer to a female student or simply the lover of a male student. But at least women can study here relatively easily. And Marie goes about doing just that with a passion. She spends as much time as she can in lecture halls, laboratories, and libraries. She dedicates the nights to her books, and attends lectures given by the legendary scientist Henri Poincaré. Once again, she overreaches herself, eventually collapsing in the library. Bronya takes her exhausted and undernourished sister in, and feeds her up with meat and potatoes until she regains her strength. No sooner has she recovered than she rushes back to her books, eventually graduating at the top of her class once again.

What now for her? Women are allowed to study, but few men are prepared to work alongside them as fellow researchers. Marie can count herself lucky to have won a scholarship to study the magnetic properties of different kinds of steel. When she has trouble getting to grips with the laboratory equipment, an acquaintance introduces her to an expert on magnetism. A shy, thoughtful man, Pierre Curie looks younger than his thirty-five years. He teaches Marie how to use an electrometer—a device he helped develop. Although Marie has vowed never to fall in love again after her heartache over Casimir, her resolve weakens, and she and Pierre become a couple.

However, the magnetic properties of steel are not Marie's scientific calling; there are far more interesting phenomena to investigate. Working in Würzburg, in southern Germany, Wilhelm Conrad Röntgen has just discovered a mysterious new kind of ray, quite by accident. Dubbed X-rays, they passed through his hand when he held it in front of an electron tube. Around the turn of the year 1896, Röntgen circulated an image of the bone structure of his wife's hand, complete with wedding ring, among his fellow scientists. It's a sensation the likes of which nobody had ever seen before. A scientific and social hype arises around Röntgen's X-ray images.

In Paris in the same year, Henri Becquerel discovers—again, quite by accident—a new kind of radiation that he names '*rayons uraniques*', 'uranic rays', after finding that they emanate from a sample of uranium placed on a photographic plate and left in a drawer. However, that's the full extent of Becquerel's discoveries concerning these new rays. He can't explain their origin. He suspects and hopes that they are somehow connected to the phenomenon of phosphorescence, which has been a favourite topic of research for him and his predecessors for generations. His rays cause much less of a sensation than Röntgen's, and his blurry images pale beside the X-ray radiographs that now adorn the front pages of newspapers and attract crowds at fairs and carnivals.

Marie Curie, on the other hand, is fascinated by Becquerel's discovery. She realises that the few experiments performed by Becquerel, who's not exactly known as a scientific workaholic, are far from enough to get to the bottom of this new effect. So she develops a new procedure to measure these uranium rays, based on the principles of Pierre's electrometers. And she dares to contradict the mighty Becquerel, describing the rays as '*radioactifs*' rather than '*uraniques*', since she is convinced that they can also be given off by elements other than uranium. In order to prove this, she sets about trying to isolate other radioactive elements—a process that will lead to the discovery of two such elements over the next two years: polonium and radium.

Furthermore, Marie claims that this 'inexplicable uranium radiation is a property of the atom', as she wrote in 1898—a provocative statement that challenges current scientific thinking. Scientists have failed to make much headway when it comes to the atom. There are simply too many different theories about it. There are the atoms of the chemists—the indivisible and immutable building blocks of matter that escape their bonds during chemical reactions and recombine with each other. Now, as of relatively recently, there are also the atoms of the physicists, shooting through the vacuum like tiny billiard balls, colliding to generate heat and pressure in gases. And then there are the atoms of the philosophers, considered the eternal basic building blocks of the cosmos since the times of Democritus. But there is no unifying theory linking these different concepts, except that they are all referred to as 'atoms'. And now, here's Marie Curie, claiming that something is going on *inside* these atoms.

How can that be possible? How might the mechanism causing atoms to emit radiation work? Experiments have shown that it's apparently unaffected by chemical processes, or by light or temperature, or by electric or magnetic fields. What is its cause? Marie Curie has an outrageous hunch: nothing. The process that causes radiation begins of its own accord—spontaneously. In a paper published for the International Congress of Physics on the occasion of the Universal Exposition in Paris in 1900, she writes a portentous sentence: 'The spontaneity of the radiation is an enigma, a subject of profound astonishment.' Radioactive radiation comes about of its own accord; it has no cause. This claim of Curie's shakes one of the very foundations of physics: the principle of causality. She even considers discarding the principle of the conservation of energy—the iron-clad rule of physics that says energy can neither be created out of nothing, nor destroyed. The man who casts light on Curie's puzzle is the physicist Ernest Rutherford. He develops a theory of 'radioactive change', describing how, when an atom emits radioactive radiation, it changes from one chemical element into another. With that, another of the dogmatic pillars of science begins to wobble. Such a transformation is thought to be impossible, nothing but the ravings of alchemists and quacks. Even Curie baulks at Rutherford's theory for a considerable time, but eventually both turn out to be right: Curie is correct about the spontaneous nature of radioactivity, and Rutherford is right about radioactive change. What is no longer right is the old theory of physics.

The Curies set up their laboratory in a converted shed at the École supérieure de physique et de chimie industrielles, in the Latin Quarter, the academic district of the French capital. Wind whistles through the cracks. The floor never gets completely dry. The shed had previously been used as a place for students to dissect corpses—until they moved to somewhere more salubrious. Now, the dissection tables have made way for strange devices: glass flasks, electric cables and vacuum pumps, scales, prisms and batteries, gas burners and crucibles. 'A cross between a stable and a potato cellar,' is how the Baltic-German chemist Wilhelm Ostwald describes the Curies' laboratory shack after his 'urgent request' to view it is granted. 'If I had not seen the worktable with the chemical apparatus, I would have thought it a practical joke.' Here, in an ambience reminiscent of an alchemist's workshop, the Curies will make some of the most important discoveries of the dawning twentieth century. Little do they know that it is in this draughty shed that they will fundamentally change the way physics explains the world around us.

Working in their shed, the Curies set out to produce a substance that until recently would also have been dismissed by their fellow scientists as nothing but hocus-pocus: pure radium. But, not being magicians who can conjure it out of thin air, they need a raw material to produce it from. After much searching, Marie comes across a mineral known as pitchblende. They will need tonnes of it, but it's not available in Paris, and the Curies have no money to pay for it, even if it were. Pierre sends out requests across Europe, and hears of the existence of an ore mine in a place called Joachimsthal, deep in the forests of Bohemia, where metal for the famous thaler coins (which eventually gave their name to the dollar) is mined. He learns that the mine produces large amounts of pitchblende as waste material, and manages to persuade the mine's director to let him have ten tonnes of the mineral. The transportation costs are met by Baron Edmond James de Rothschild, the super-rich son of the famous banker who has always been far more interested in art, science, and horseracing than in the banking business of his father.

When a mountain of pitchblende is delivered to the yard outside their shed in the spring of 1899, Marie picks up a handful of the 'brown dust mixed with pine needles', and holds it to her face. Now she can get to work.

It's backbreaking labour. Marie heaves heavy buckets, pours off liquids, stirs boiling cauldrons with iron rods. The pitchblende must be rinsed with acid, alkaline salts, and thousands of litres of water.

To separate the substances in the pitchblende, the Curies develop a technique they call 'fractional crystallisation'. They repeatedly boil up the raw material and leave it to cool and crystallise. Light elements crystallise more quickly than heavy elements—a phenomenon that the Curies make use of to gradually concentrate the radium present in the pitchblende. Identifying it requires a lot of patience and precise measurement, but the couple are happy, despite the gruelling work. As they walk through the city together each night on their way home from the laboratory, they try to imagine what pure radium might look like. As their radium mixture becomes ever purer, the glowing light that emanates from the flasks at night becomes ever brighter. In the summer of 1902, their work finally pays off, and they have one-tenth of a gram of radium. Marie's measurements of the element's atomic weight place it at number 88 on the periodic table.

But one member of the family is not so happy—Irène, the Curies' daughter, who was born two years before her parents set up their shed-laboratory. Irène sees little of her mama and papa, and they're always exhausted when they do come home from work. Irène is looked after by her grandfather Eugène, and she shows all the signs of a child with separation anxiety. Whenever her mama, Marie, makes to leave the room, Irène clings to her skirts, crying. One day, she asks her grandfather why her mother is away from home so much. Her grandfather takes her by the hand and brings her to the laboratory shed. Irène is appalled at the sight of 'this sad, sad place'. Once again, a daughter who misses her mother. Three decades later, now known as Irène Joliot-Curie, she will win a Nobel Prize for her own research into radioactivity, becoming only the second woman to do so, after her mother. Irène's own daughter, Hélène, will also become a nuclear physicist.

Little does Marie Curie suspect, on that July evening in the Boulevard Kellermann, that tragedy is about to befall her family. She has had a new dress made for the party, of black cloth, all the better to hide the stains from the laboratory. And the bulge in her belly. A few weeks later, she takes a bicycle ride with Pierre. She and her husband love to ride through the countryside; even their honeymoon was spent in the bike saddle. But Marie is now five months pregnant, and the jolts from cycling on the bumpy dirt roads prove too much for her body. She suffers a miscarriage. To escape her grief, Marie buries herself ever more deeply in her work, until she finally suffers another breakdown. She's unable to travel to Sweden to accept the Nobel Prize awarded to her and Pierre, together with Henri Becquerel, for their discovery of radioactivity, leaving the stage to the vainglorious Becquerel, who appears wearing a green-and-gold embroidered frockcoat, medals on his chest, and a sabre hanging from his belt.

When Marie takes Pierre's arm and walks with him through the salon door and out into the warm night on the summer evening of her doctoral graduation party, the guests all raise their glasses to her. The couple step out of the light to catch a brief moment alone. Under a starlit sky, Pierre reaches into his waistcoat pocket and pulls out a glass vial of radium bromide. Its glow illuminates their happy, alcohol-flushed faces, but also the burnt and cracked skin on Pierre's finger—the first signs of radiation sickness, which would one day kill Marie, and a first indication of the power hidden in the knowledge they are seeking.

BERLIN, 1900

An act of desperation

The seventh of October 1900 is a Sunday, and it's promising to be a boring one. Max and Marie Planck have invited another couple, their neighbours Heinrich and Marie Rubens, to afternoon tea at their apartment in Berlin's leafy, upper-middle-class district of Grunewald. Rubens is the professor of experimental physics at the University of Berlin, while Planck is the professor of theoretical physics. Much to their wives' annoyance, the two men can't help talking about work. Rubens is describing his latest set of measurements at the Imperial Institute for Physics and Technology, and the fact that the results he and his colleagues have recorded contradict all previously accepted formulae. He talks of wavelengths, energy densities, linearity, and proportionality. Inside Planck's head, the jigsaw pieces he's been shifting around in his mind for years begin to form a new pattern. That evening, long after the guests have left, Planck sits at his desk and sets down on paper what has formed in his mind: a formula for radiation that accurately explains all the measured data, the equation that Planck and many others have spent years searching for. Sometime around midnight, Marie Planck is woken by the sound of her husband playing Ludwig van Beethoven's *Ode to Joy* on the piano. It's his way of expressing his elation. Even before the night is out, he writes the formula on a postcard and sends it to Rubens.

'I've made a discovery as important as Newton's,' forty-two-year-old Max Planck announces to his seven-year-old son, Erwin, as they take a morning walk through Berlin's Grunewald forest. And he's not exaggerating.

Planck is anything but a born revolutionary. Indeed, he's the epitome of a Prussian official, always very properly dressed in his dark suit, starched shirt, stiff collar, and black bowtie, a pince-nez perched on his nose to correct his short-sightedness. His eyes are piercing beneath the high dome of his bald head, under which sits a cautious mind. He describes himself as having a 'peaceable nature'. 'My maxim,' he confides to one student, 'is always to consider every step carefully in advance, but once I've decided it's justified, to let nothing stand in my way.' This how he approaches new ideas and reconciles them with his extremely conservative worldview. 'It's inconceivable that this should be the man to spark a revolution,' says one of Planck's students. Both he and the rest of the world are soon to learn how wrong they are.

Max Karl Ernst Ludwig Planck was born in 1858, in the northern German port city of Kiel, which was controlled by the Kingdom of Denmark at the time. He comes from a long line of academics. Both his paternal grandfather and great-grandfather were eminent theologians, and his uncle Gottlieb Planck is one of the authors of Germany's Civil Code—a groundbreaking book of law that is still largely in force today and that became the template for the civil legislature of many countries around the world. Max Planck's father, Johann Julius Wilhelm Planck, also a law professor, was awarded the Knight's Cross by Bavaria's King Ludwig II in 1870, earning him the right to call himself 'von Planck' in the aristocratic style. All members of the Planck family are loyal patriots with a great respect for the law, both religious and secular. And, growing up as part of that family, Max is no exception.

Just after Max Planck turns nine, the family moves to Munich, occupying a large apartment at Briennerstrasse 33. Max's father is now professor of civil procedural law at Munich's Ludwig Maximilian University, and Max takes up his studies in the sixth grade at the Maximiliansgymnasium—commonly known as the 'Max'—just after the prestigious high school moves into new premises in a former nunnery at Ludwigstrasse 14.

He is not necessarily the best of the sixty-five students in his class, but he is very disciplined. He always receives top marks for 'moral conduct' and 'diligence', and is gifted in the most important abilities required by the Prussian education system—built, as it is, on the principle of learning large amounts of material by rote. One school report describes Max as having every chance of becoming 'something decent' in life. It goes on to say he is 'a favourite with both his teachers and his classmates, possessing a very ordered and logical mind beyond his young years'. The teenage Max Planck is not drawn to the beer cellars of Munich, but to its opera houses and concert halls. With a great gift for music, he displays perfect pitch from an early age. He plays both violin and piano, and often sings female solo soprano parts as a church chorister. He also plays the organ in church on Sundays, and composes songs himself—even writing an entire operetta. His *Love in the Forest* is performed at a festival of the university's prestigious Academic Choral Society.

After sailing through the school-leaving exam at the age of sixteen, Planck considers a career as a concert pianist. But on asking a professor about the prospects of studying music, he receives a rather harsh answer: 'If you even need to ask that question, go and study something else!' Perhaps classics is the subject for him? Max is undecided. His father arranges a meeting with the physics professor, Philipp von Jolly, who does his best to dissuade Max from studying the subject. He describes physics at the time as 'a highly sophisticated, almost fully developed science which will soon have achieved its final, stable form, now that it has been crowned, so to speak, by the discovery of the principle of the conservation of energy. A little dust mote or bubble may remain in some corner somewhere, waiting to be examined or classified, but the system as a whole is quite established, and theoretical physics is approaching the level of perfection already reached by the field of geometry centuries ago.'

Jolly is not alone in holding this view. Up until the dawn of the twentieth century, physicists were confident they would soon bring their discipline to the final point of perfection. 'The more important fundamental laws and facts of physical science have all been discovered,' declared the American physicist Albert Michelson in 1899. 'And these are now so firmly established that the possibility of their ever being supplanted in consequence of new discoveries is exceedingly remote. Our future discoveries must be looked for in the sixth place of decimals.'

James Clerk Maxwell, the founding father of the classical theory of electromagnetism, warned against such complacency as early as 1871: '[The] characteristic of modern experiments—that they consist principally of measurements—is so prominent, that the opinion seems to have got abroad, that in a few years all the great physical constants will have been approximately estimated, and that the only occupation which will then be left to men of science will be to carry on these measurements to another place of decimals.' Maxwell stressed that the real reward for 'the labour of careful measurement' was not greater precision, but rather 'the discovery of new fields of research' and 'the development of new scientific ideas'. Subsequent scientific developments were to prove Maxwell's prophesy right.

Jolly can't know at the time that it is this historical error that will condemn him to occupy only a minor place in the history of physics. Nor can he know that it is the sixteen-year-old Max Planck sitting before him now who will be the one to expose this error. Planck himself also has no idea of this at the time. Measuring to a few more decimal places and refining calculations—that doesn't sound so bad to him. It's certainly more promising than the music professor's answer, in any case. So Max Planck enrols at the university to study mathematics and natural sciences, beginning in the winter semester of 1874–75.

Once at the university in Munich, Planck finds he is as bored as Philipp von Jolly had predicted. Jolly's research projects include one to make the most precise measurement to date of the specific weight of liquid ammonia, using a homemade spring balance, and one to verify Newton's law of gravity, using a lead ball with a weight of 5775.2 kilograms and a diameter of almost one metre—anything but revolutionary science.

Planck lasts three years at the university's faculty of physics before the boredom finally gets to him and he transfers to Berlin. The university there is an important centre for the study of physics, boasting such luminary teachers as Gustav Kirchhoff and Hermann von Helmholtz.

After the North German Confederation's victory over France in the Franco–Prussian War of 1870–71 and the emergence of a united Germany, Berlin becomes the capital of a new and powerful European nation. War reparations from France fund the growth of a metropolis at the confluence of the Havel and the Spree to rival both Paris and London. Between 1871 and 1900, Berlin's population grows from 865,000 to more than two million, making it the third-largest city in Europe. Those numbers are swelled by many migrants from the east—principally Jewish people fleeing the pogroms in czarist Russia.

Berlin's ambitions to become an important European metropolis go hand in hand with a desire to turn the city's university into one of the best on the continent. Hermann von Helmholtz, the country's most eminent physicist, is summoned from Heidelberg. Helmholtz is a polymath in the old style: a qualified surgeon and a celebrated physiologist. As the inventor of the ophthalmoscope, he has greatly advanced scientific understanding of the functioning of the human eye.

Few scientists of the time have such a broad horizon as Helmholtz. And the fifty-year-old scholar is aware of his own worth. The salary he negotiates is many times higher than the norm, and he is provided with a magnificent new physics institute, which is still under construction when Max Planck arrives in Berlin in 1877 and attends his first lectures in the university's main building—a former palace on the grand central boulevard Unter den Linden, opposite the Opera House. For Planck, it feels as if he has emerged from a cramped room into a great hall.

But even in great halls, things can be boring. Planck finds Kirchhoff's lectures, read aloud from a script in a notebook, 'dry and monotonous', while those given by Helmholtz are ill-prepared, poorly presented, and riddled with miscalculations. Planck, still the eager high school student at heart, resorts to self-directed study. He reads the works of Rudolf Clausius on thermodynamics and entropy—the new physical measure of disorder, and a first step towards a scientific revolution.

Planck passes his final university exams in physics and mathematics at the age of twenty. A year later, he presents his doctoral thesis, *On the Second Law of Thermodynamics*. And another year after that, he delivers his dissertation with the title *Equilibrium States of Isotropic Bodies at Different Temperatures*, which earns him his habilitation—the post-doctoral qualification required to take up a university professorship in Germany. He is awarded the two titles with the honours '*summa cum laude*' and 'highly satisfactory', respectively. Planck appears to have an exemplary academic career ahead of him.

Returning to Munich, Planck becomes a lecturer at Ludwig Maximilian University and moves back in with his parents, where he enjoys 'the nicest and most comfortable life imaginable'. However, that soon comes to an end when he's offered a professorship in the city of his birth, Kiel. The salary of 2,000 marks a year is just enough to start a family of his own. All he needs now is to find a suitable wife. He soon marries Marie Merck, the sister of an old schoolfriend from a wealthy banking family. The couple have three children within the next two years.

Just as Max Planck is settling into life as a family man, fate steps in again. The ailing Gustav Kirchhoff passes away, leaving the chair of mathematical physics at Berlin's Friedrich Wilhelm University vacant. The appointments committee casts around for a candidate 'of solid scientific authority who is in the vigorous prime of manhood'. Both Ludwig Boltzmann, the founder of statistical mechanics, and Heinrich Hertz, the discoverer of electromagnetic waves, had been offered the post but declined it. Max Planck is the third choice, but, at just thirty years of age, is he mature enough to take up one of the country's most important professorships? Some members of the various panels of Berlin physicists, whose average age is often around sixty, doubt it. However, after his former teacher, Hermann von Helmholtz, steps in to support his nomination, Planck is awarded a position, albeit initially only as an associate professor.

So Planck has to prove himself worthy of the position previously occupied by his former teacher, working alongside his other principal teacher, Hermann von Helmholtz. Planck decides to tackle a problem that Kirchhoff left unsolved, known as the blackbody problem.

As potters and blacksmiths have known for centuries, all objects, no matter what they are made of, glow in a sequence of characteristic colours as their temperature rises when they are heated. If you hold a metal poker in a furnace, it initially begins to glow faintly with a dark-red colour, which gradually becomes a brighter cherry-red, before changing to yellow as the iron gets hotter. The higher its temperature rises, the brighter and whiter the poker will glow, until it gradually begins to take on a bluish hue. This characteristic sequence of colours is always the same, no matter where or what; from the red of a glimmering ember in the hearth, via the yellow of the Sun, to the bluey whiteness of molten steel.

Experimental physicists had measured the spectra of the emitted radiation again and again. With improved thermometers and photographic plates, they had discovered that the palette of colours extends beyond the visible range—into the infrared at the cooler end and the ultraviolet at the hotter end. They were advancing their knowledge by one decimal place at a time.

What was needed was a formula to describe accurately the correlation between the temperature and the frequency (that is, the colour) of the light emitted. This was the blackbody problem, so named because it is based on a theoretical, idealised physical body that absorbs any and all electromagnetic radiation incident upon it. The blackbody problem had been formulated scientifically in 1859 by Gustav Kirchhoff, who was a professor in Heidelberg at the time, and an authority on the spectrum analysis of mineral water. But he and his fellow scientists had repeatedly failed to find a formula for blackbody radiation. Wilhelm Wien had proposed a formula that predicted behaviour at high frequencies reasonably well, and James Jean developed a different equation to describe blackbody radiation at greater wavelengths. But the two laws were mutually incompatible, breaking down at each end of the spectrum, respectively.

The blackbody problem was not the only issue confronting physicists. X-rays, radioactivity, and electrons had all been discovered recently, and a dispute over the existence of the atom was raging. Compared to all that, the blackbody problem seemed trifling. But that's precisely why it niggled at the experts of the day.

And solving the problem was more than just an exercise in mental gymnastics. Indeed, it was a matter of national importance. The German Empire had been proclaimed in 1871, and the young nation now hoped, by solving the blackbody problem, to gain an advantage over Great Britain and the United States in the internationally competitive lighting-technology industry. From the point of view of physics, a lightbulb filament was no different to a glowing hot poker. Thomas Edison had received a patent for his incandescent lightbulb in 1880. It was a far better lighting solution than the gas lamps commonly used at the time, and it launched an international race to control the lighting market. German companies attempted to get the better of their American and British rivals by producing more efficient bulbs.

The young German Empire was in a good position in the international race for dominance in electronics and electrical engineering. Werner von Siemens had invented the dynamo just ten years earlier. The government set up the Imperial Physical Technical Institute on the outskirts of Berlin in 1887, with Siemens' support. Its mission included a program to study blackbody radiation, with the aim of making German lightbulbs the best in the world.

When Max Planck succeeds Kirchhoff at the university in Berlin, he has to prove he has what it takes to fill his predecessor's shoes. He has to hold his own in the academic hustle and bustle of the capital city's university: tutoring and examining hundreds of students, writing reports, attending meetings. His lectures are as dull and uninspiring as those of his predecessor. 'For all their extraordinary clarity, somewhat impersonal, almost to the point of being dry,' complains one student, a young woman by the name of Lise Meitner. 'Planck is not exactly a barrel of laughs,' another student quips.

From 1894 on, Planck devotes all his available research time to the blackbody problem that Kirchhoff left unsolved. He's fascinated by the fact that 'black cavity radiation' appears to be 'something absolute', 'and as I had always looked upon the search for the absolute as the noblest and most worthwhile task of science, I eagerly set to work on it'. He attacks the blackbody problem with the weapons of pure theoretician: paper, a pencil, and his intellect. However, no sooner has he come up with the long-sought-after formula that he scribbles down so joyfully on that Sunday night than he faces the next challenge: he doesn't understand his own discovery. Two weeks later, on 19 October, when he rises to speak after a talk by Ferdinand Kurlbaum at the German Physical Society's Friday colloquium, he can report little more than the formula itself.

The difficult part still lies ahead of Planck: interpreting and substantiating the law he has discovered by intuition. In physics, it's not only important to come to the correct answer, but also to understand *why* that answer is correct. In the weeks following his lucky discovery, Planck tries to derive it using the principles of physics. He's an old-school physicist, with little time for new-fangled methods such as Ludwig Boltzmann's statistical mechanics, or even atomic theory. But the concepts of classical physics do not allow him to understand his own equation. What is the meaning of the mysterious constant called *h*, which he put down on paper so casually on that fateful night? Its value is tiny, at just 0.000000000000000000000000000655 (a figure with twenty-six zeros after the decimal point). But, try as he might, he can't reduce this *h* to zero.

In what he later calls 'an act of desperation', Planck forces himself to accept the fact that the blackbody is indeed made of atoms. Using Boltzmann's statistical methods, which he had formerly rejected out of hand, he manages to derive his formula. But this desperate act has a further, rather strange repercussion: his conclusion 'that energy is forced at the outset to remain together in certain quanta'. First atoms, now 'quanta'! Planck hopes this theoretical nightmare will soon disappear and that only his formula will remain. He describes the concept of quanta as 'purely a formal assumption, and I really did not give it much thought except that, no matter what the cost, I must bring about a positive result'. Just a mathematical trick. Nothing that's going to rock anybody's worldview. Yet.

On 14 December 1894, at five o'clock in the afternoon, Planck once again speaks at the Physical Society's Friday colloquium. His lecture is titled 'On the Theory of the Energy Distribution Law of the Normal Spectrum'. The audience on the wooden benches includes researchers such as Heinrich Rubens, Otto Lummer, and Ernst Pringsheim. 'Gentlemen!' Planck greets them. Then, with typically long, convoluted sentences, he begins:

> When, some weeks ago, I had the honour to draw your attention to a new formula which seemed to me to be suited to express the law of the distribution of radiation energy over the whole range of the normal spectrum, I mentioned that, in my opinion, the usefulness of this equation was not based only on the apparently close agreement of the few numbers which I could then communicate, with the available experimental data (a direct conformation of which for very long wavelengths has in the meantime been provided by Messrs. Rubens and Kurlbaum), but mainly on the simple structure of the formula and especially on the fact that it gave a very simple logarithmic expression for the dependence of the entropy of an irradiated monochromatic vibrating resonator on its vibrational energy, which seemed to promise in any case the possibility of a general interpretation much rather than other equations which have been proposed, apart from Wien's formula, which, however, was not confirmed by experiment.

In other words, he has already announced the formula, and now he can explain it. He soon comes to the key step in his derivation: 'We consider, however—this is the most essential point of the whole calculation—energy to be composed of a very definite number of equal parts and use thereto the constant of nature *h = 6.55×10–27 ergsec.*' Quanta have just made their first appearance in world science, and nobody even noticed. There is warm applause from the audience.

Neither Planck nor anyone else in that audience realises that later physicists will refer to that afternoon as 'the birth of quantum physics'. For many years after, Planck and other physicists, such as Lord Rayleigh and James Jeans in England and Hendrik Antoon Lorentz in Leiden in the Netherlands, will try to disprove the existence of quanta. They still believe energy is a continuum and that it travels through the ether. They still believe in Newton and Maxwell. But all that will fall, while quantum theory is here to stay.

BERN, 1905

The patent serf

Bern, Switzerland, 17 March 1905. The mediaeval clock tower, the Zytglogge, is about to strike 8.00 am. A young man in a checked suit hurtles down the steep, narrow stairway from the second floor of the house at Kramgasse 49, along the cobbled streets and through the covered mediaeval arcades. In his hand, he clutches an envelope. Passers-by may be astonished to notice his footwear: worn green slippers, embroidered with flowers. But the young man pays no heed to their surprised glances. He needs to get to the post office urgently. The contents of that envelope he's carrying are going to change the world. The young man's name is Albert Einstein.

Einstein turned twenty-six just three days earlier. He became a father ten months ago. He lives in a one-bedroom flat on the second floor with his wife, Mileva, and their baby son, Hans Albert.

Einstein works at the Swiss Patent Office as a 'technical clerk, 3rd class'. It's not his dream job, but he's glad to have found any employment at all, after his application for doctoral studies was rejected, with no assistantship at the university, the difficult birth of his son, and a complex, laborious application process at the Patent Office. Einstein has to take on extra work for a while as a private tutor to pay the rent and provide for his wife and son. He teaches architects, engineers, and long-term students in physics and mathematics. One of his pupils, who was from the French-speaking part of Switzerland, noted in his copybook: 'His short skull looks extremely broad. His complexion is of a dull light brown colour. Above his large, sensual mouth sprouts a slender black moustache. His nose is slightly aquiline in form. There is a deep, soft gleam to his dark-brown eyes. His voice is engaging, like the sound of a resonating cello. He speaks accurate French with a slight foreign accent.' At the same time, Einstein attends lectures in pathology at the University of Bern. The physics lectures are just too boring. His attempt to become a university academic himself is thwarted when the university rejects his application for habilitation—the post-doctoral academic rank required for teaching at a university—on the grounds that his achievements are not sufficient to exempt him, who does not even have a doctorate, from the requirement to present a habilitation thesis. Einstein calls the university 'a pigsty. I shan't study there.' This is the end of Einstein's first, failed attempt to become a 'great professor'.

The past few years have been difficult for Einstein for many reasons. In 1896, when he is seventeen years old, he finally enrols to study at the Federal Technical University in Zurich, having initially failed the entrance exam, and having had to take a circuitous route to acceptance by resitting his school leaving qualification, this time in the Swiss system. At exactly this time, Einstein's father's company goes bust, leaving the young student with no financial support in the largest and wealthiest city in Switzerland, the banking and business metropolis of Zurich. Relatives in Italy send him 100 francs a month to help him make ends meet. He scrapes through his physics degree, even receiving a note of censure alongside a bad grade in his 'practical physics for beginners' course. He often skips classes without permission, preferring to study by himself at home, reading the classic works on electromagnetism by James Clerk Maxwell and Heinrich Hertz, as well as the newer works by Ludwig Boltzmann, Hermann von Helmholtz, and Ernst Mach.

Einstein is particularly drawn to Mach, the Vienna-based physicist who champions a new philosophy of science and a fundamental rethinking of physics, free from unsubstantiated hypotheses and metaphysical speculations. In Mach's view, everything must be based exclusively on directly observable phenomena. Physical concepts such as velocity, force, and energy must be rooted in sensory experience. Ideas such as absolute space and absolute time, part of the scientific dogma since Newton and seen as the extrasensory prerequisites for sensory experience since Kant, are part of the metaphysical junk that Mach means to clear out. There is no such thing as absolute time. There are just the hands and bells of the Zytglogge.

Whenever he's asked whether atoms really exist, Mach likes to respond, 'Well, have you ever seen one?' Of course, he can be safe in the assumption that the answer will be 'no'. But that is about to change. The 'uranic rays' observed and investigated by Henri Becquerel and the Curies are evidence of the existence of atoms, and Einstein is not the sort of person to ignore what he can see with his own eyes.

Einstein has to accept that he is 'a mediocre student', and graduates in fourth place out of a class of five. Their physics professor, Heinrich Friedrich Weber, awards assistantships to all of them — except Einstein. His two attempts to gain a doctorate also fail, as the professors reject his doctoral theses. Later in life, Einstein himself would call these dissertations 'my first two worthless works'.

Einstein's girlfriend, Serbian Mileva Marić, is one of the first women to study physics. But she fails her final exam, gets pregnant by Einstein, retakes the exam but fails it again, and gives birth to a daughter called Lieserl. Mileva and Albert keep the birth of their illegitimate daughter a secret from family and friends, and give her up for adoption. Einstein, having already moved to Bern, never even sets eyes on his daughter. Mileva joins him, and the couple marry — against the will of Einstein's mother. Not exactly what were called 'orderly circumstances' at the time.

When Einstein finally lands the job at the Patent Office, his financial worries are over. His 'pretty salary' of 3,500 francs a year is enough to finance a middle-class family life. But money worries are soon replaced by stress at work. Einstein has to appear at 'the Office' every weekday morning at 8.00 am, on the second floor next to the Post and Telegraph Directorate, where he spends eight hours checking patents. After work, he has at least one private student a day. At first, he even has to receive extra training from his boss, as his knowledge of engineering and technical drawing leaves so much to be desired.

Cut off from the centres of physics research as he is, no one would blame Einstein for concentrating on his career in the Swiss civil service. However, such academic isolation is precisely the environment he thrives in. He needs this separation from the established world of physics to form his own ideas, but he is far from being the solitary genius, the 'lone wolf' he likes to see himself as. Ever since she and Einstein met as students in Zurich, Mileva has always been a source of intelligent, congenial, and like-minded conversation, and at times her insights can barely be separated from his.

Calling themselves the 'Olympia Academy', Einstein and a group of good friends meet regularly to talk and bluster about physics and philosophy, free of the restraints of academia. Einstein's invitations to sessions of the 'academy', which can't be missed without good reason, are signed 'Sir Albert, Knight of the Coccyx'—a jocular reference to his ability and willingness to sit and listen for so long.

Einstein also regularly attends the fortnightly evening sessions of the Bern 'Natural History Society' in the function rooms of the Storchen Hotel. He hears scholarly talks by retired professors, schoolteachers, doctors, and pharmacists. On 5 December 1903, Einstein gives a talk himself. The topic is his 'theory of electromagnetic waves'. Later, this will become known as his 'theory of relativity'. 'A scholarly paper only exists in my head so far,' Einstein tells the audience, before the society moves on to the topic of veterinary medicine.

When Einstein reads Max Planck's 1900 paper on the blackbody problem, he is the first to realise the implications of this discovery. 'It was as if the ground had been pulled from under one's feet, with no solid ground left to build on.' If light is made up of 'quanta', as suggested by Planck's work, how could he continue to believe Maxwell's wave theory of light? Einstein decides to venture out into unknown territory and to take Planck at his word.

Since James Clerk Maxwell's work more than half a century before, scientists have considered light to be a form of wave energy. But in solving the blackbody problem, Planck was forced to go against his own scientific intuition and accept the idea that energy is absorbed and emitted in discrete lumps. Energy doesn't flow evenly, but is emitted or absorbed in very definite, tiny units—quanta. But he continues to believe, like all physicists, that electromagnetic radiation consists of constantly oscillating waves. How is anything else possible? These troublesome lumps of energy must arise somehow when matter and radiation interact. Burning with the revolutionary spirit that Planck so lacked, Einstein claims that light, and indeed all electromagnetic radiation, consists not of waves, but of particle-like quanta.

This is the bold statement contained in the manuscript inside the envelope that Einstein is taking to the post office on his way to work on that morning of 17 March 1905. The envelope is addressed to the publisher of the *Annals of Physics*, the world's leading physics journal at the time. His paper is entitled 'On a Heuristic Point of View Concerning the Production and Transformation of Light'. Einstein is well aware that his proposal is even more radical than that of Max Planck. Light as a stream or particles—that's tantamount to heresy.

For the next twenty years, almost no one other than Einstein will believe in the particle theory of light. From the outset, Einstein realises that it will be a difficult struggle. By including the word 'heuristic' in the title of his paper, he admits that he doesn't consider his 'view' to be a rigorously developed theory, but simply a working hypothesis, an aid to understanding more about light's puzzling behaviour. This makes it easier for his fellow physicists to appreciate and accept his view at all. It's a signpost pointing towards a new theory of light. But even that proves too much for Einstein's fellow physicists, who are simply incapable of thinking of light in any way other than that set out by Maxwell. It is not until decades later that they are finally able to follow Einstein on the journey into new dimensions that he embarked on from the desk of his office in 1905.

And that's just the beginning of the bombshells that this Bern-based patent clerk, 3rd class, will drop on the world of physics in the year 1905. In May, Conrad Habicht receives a letter from his friend Albert Einstein. Habicht moved away from Bern a few months earlier to take up a position as mathematics master at a village school in rural Switzerland. The letter was clearly written in a hurry, with a nervous hand and full of ink blots and corrections. Einstein hasn't even thought to write the date on it. It begins with a few insults, calling Habicht a 'frozen whale' and a 'dried and bottled-up bit of soul' for whom he felt 'seventy per cent anger and thirty per cent pity'. This is Einstein's way of showing affection. He misses Habicht and their sessions together at the Olympia Academy.

Then Einstein goes on to promise to send his friend four papers that he hopes will be published before the year is out. The first deals with light quanta. The second, which forms his doctoral thesis, describes a new way of measuring the size of atoms. In the third paper, Einstein explains Brownian motion—the erratic dance of particles such as pollen in a fluid, which has been a puzzle to scientists for the past eight decades. 'The fourth paper exists in concept form,' writes Einstein, 'and is on the electrodynamics of a moving body using a modification of the theory of space and time.' As a hobby physicist, Einstein has achieved the aim first instilled in him by reading Ernst Mach: reinventing space and time. Max Planck, who reviews the papers for the *Annals of Physics*, gives the new theory a name that is to become virtually synonymous with the name of Albert Einstein: 'the theory of relativity'.

However, it is not the theory of relativity that Einstein labels 'highly revolutionary' in his letter to Habicht, but the quantum theory of light. This is the only time he refers to one of his works as 'revolutionary'. The peer reviewer of the paper, Planck, who himself came up with the concept of quanta, still regards the idea as nothing more than a temporary mathematical aid, and is absolutely opposed to Einstein's particle theory of light. But he approves the paper for publishing. 'Who is this amateur physicist from Bern who's suddenly spewing out all these fantastical and reckless theories?' Planck wonders.

The works that Einstein lists in his letter to Habicht alone are easily sufficient to earn Einstein an eternal place of honour in the history of science. Einstein produces them in his spare time, within the space of just a few months. Never before has the world seen such an outburst of creativity in a scientist. And then he goes on to write a fifth paper, which he doesn't mention in his letter to Habicht. It contains the equation $E=mc^2$.

Einstein receives his doctorate from the University of Zurich in January 1906. This leads to a promotion at the patent office, to patent

Albert Einstein in his study in Berlin, 1921

clerk, or, as he himself puts it, 'patent serf, 2nd class'. His salary rises to 3,800 francs a year. In early 1907, Einstein writes in a letter to a friend, 'I'm doing well; I'm a respectable Swiss ink-shitter with a good salary. I'm also still riding that mathematical-physical hobbyhorse of mine and scraping away at the violin—both within the narrow limits for such superfluous pursuits imposed by my two-year-old little boy.'

PARIS, 1906

The decline and fall of Pierre Curie

Marie and Pierre Curie are now stars. Newspapers publish celebrity 'at-home-with' features about them, and gush over their 'physics lab love story'. At the same time, interest in the new element radium has become a global hype. Radium is said to be able to cure cancer, whiten teeth, and increase your libido. The glitterati are delighted by radium light effects at their parties, while dancing girls daubed with radium paint entertain them. Radium factories spring up around the world and have to compete for access to supplies of pitchblende. The American steel magnate and athlete Eben Byers famously drinks a bottle of radium water every day to give him what he calls a 'toned-up feeling'. He is to suffer an agonising death from cancer of the jaw.

The Curies also investigate the physiological effects of radium. They attach rubber-covered pellets of radium salts to their skin, and record the changes caused by radiation. First, redness, then blistering and ulceration. For one experiment, Pierre wraps a weakly radioactive sample round his arm for ten hours. The ensuing wound takes four months to heal. Marie and Pierre are beginning to show the first signs of what will later prove to be radiation sickness. The skin on their hands is cracked and sore. Pierre can barely sleep anymore, due to the severe pain in his bones. Radioactivity is absorbed by the clothes they wear, the paper they write on. A century later, they still cause a Geiger counter to click.

One day in 1906, Pierre leaves the house after an argument with

Marie. He limps through the streets, angry and in pain. At a busy intersection of the rue Dauphine, he slips and falls under a horse-drawn cart. One of the vehicle's back wheels crushes his skull, killing him instantly. Devastated by the loss of her husband, Marie moves house to be closer to his grave. Not a single photo of her taken after Pierre's death shows her smiling. The same year, she becomes the first woman to teach at the Sorbonne, and in 1908, two years after the tragedy, Marie is appointed to Pierre's former professorship in physics. In 1911, she wins a second Nobel Prize—this time in chemistry, for the isolation of radium. This time, there is no party.

BERLIN, 1909

The end of the flying cigars

Berlin, summer 1909. Berliners are heading in their thousands to the Prussian Army parade ground called Tempelhofer Feld. They arrive by bicycle, by underground railway, or on foot. The people of this time love nothing better than to witness one of the world's new technological marvels.

In the jostling crowd, few spectators take note of a pyramid-shaped wooden tower supporting a weird contraption. They don't see the flying machine that Orville Wright had dismantled, packed into crates, shipped from America to Europe, and reassembled in Berlin. The contraption on top of the wooden tower propels Wright and his flying machine into the sky, and he sets a new world record for flight altitude, of 172 metres above the ground, to the clamorous delight of the people of Berlin.

Just a few days earlier, Orville Wright was among the crowds of spectators. From his position next to the Kaiser on the stands, he admired the pride of Germany's aviation efforts: the zeppelins. Count Ferdinand von Zeppelin himself was piloting one of his giant cigars. He flew higher and further than Wright, but his airship appeared slow and clumsy alongside the Wright brothers' filigree flying machines. The count sedately lowered the bow of his ship in a nod to the stands and a bow to the Kaiser. Wright had applauded politely. These zeppelins belonged to the past. The Wrights' aeroplanes were the future. And the next world war will be fought in the air.

PRAGUE, 1911

Einstein says it with flowers

April 1911. Albert Einstein is settling into his new office at the German University in Prague, where he has moved from Zurich. His window looks out on the ancient trees of a beautiful park. Einstein is surprised to see only women walking in the park in the morning, while the afternoon strollers are all men. He eventually learns that the park is part of what is then known as an 'insane asylum'. 'There you see that fraction of the insane who are not working on quantum theory,' Einstein likes to tell visitors. Quantum theory seems to be threatening his own mental health. He is plagued by the idea of quanta and the dual nature of light: ideas that he brought into the world himself. Do they really exist? Eventually, he stops worrying about it. After giving a lecture on 'The Theory of Radiation and Quanta' at the first Solvay Conference in 1911, he decides to put aside this quantum madness and devote himself to problems with fewer side effects for his mental health. He needs to get away from the gloom of Prague. On a recommendation from the French mathematician Henri Poincaré, who has arrived at the equations for special relativity independently of Einstein, the Federal Technical University in Zurich offers Einstein a professorship. He returns to Zurich in July 1912 as a professor at the very institution that once refused him an assistantship.

But, once again, his stay is not long. Just one year later, Einstein greets physicists Max Planck and Walther Nernst at Zurich railway station. He knows they are here to persuade him to move to the

German capital. But he doesn't know exactly what they are offering. A salary of 12,000 marks a year, Planck tells him, which is the maximum pay for a professor in Prussia, plus an honorary gratuity of 900 marks a year from the Prussian Academy of Sciences. Einstein is impressed. But he asks for a day to think it over, as he has other offers to consider. While Planck and Nernst take a trip up Mount Rigi on the mountain railway, Einstein mulls their proposal. He has told the physicists they will know his answer by the colour of the flowers they see. Red will mean yes; white will mean no. When they return and meet Einstein, they see him holding a bunch of red flowers in his hand.

CAMBRIDGE, 1911

A Dane grows up

September 1911. A young Danish man, not yet twenty-six years old, arrives in the English university town of Cambridge. There is an air of sadness about him—with his timid eyes below thick, hanging brows, and his drooping mouth. Whenever he thinks really hard, his arms with their oversized hands hang down at his sides, and his face sags, making him look 'like an idiot', as one colleague puts it. And that's also how he sounds when he speaks, in slow, deliberate sentences.

But appearances can be deceptive. Niels Bohr is a man of huge power, both physically and mentally. He skis and skates in winter, and in summer he plays the new, highly fashionable game from England that's currently taking the continent by storm: football. Bohr is the goalkeeper for the Akademisk Boldklub, the football club founded by his father. And he's one of the most gifted scientists of his generation. He just needs to prove it—both to himself and to the rest of the world.

Bohr's career as a scientist begins with something of a false start. He has recently finished writing his dissertation on the electron theory of metals. He believes that electrons in metals behave like atoms in a gas as they 'bump' their way freely through the conductor, laden with electric charge. This model doesn't stand up particularly well, but he is awarded his doctorate anyway. No one else in Denmark at the time has enough understanding of electrons to argue against his thesis.

Bohr begins to realise that something is not right with the nineteenth-century view of electrons as tiny, charged billiard balls. He has come here to Cambridge to study under the master of the electron,

Joseph John Thomson. The fifty-five-year-old physicist, known to everyone simply as J.J., runs the famous Cavendish Laboratory, founded by James Clerk Maxwell, and is a professor at Trinity College, where Isaac Newton once taught. It is Thomson who had discovered the electron fifteen years earlier. Perhaps Thomson can help the young Dane get his dissertation published in a prestigious scientific journal?

Bohr's great ambition is to discover how atoms work. Scientists at this time know little more about atoms than that they exist. Cambridge should be just the right place for Bohr to achieve this, as Thomson has the same ambition. If only it weren't for the strange manners of the English.

Thomson is a highly respected laboratory director, but he's infamous for his clumsiness as an experimental physicist and for his awkwardness in social situations. Not long after arriving in Cambridge, Bohr has the audacity to point out to the great man Thomson a number of mistakes and inaccuracies in his book *Conduction of Electricity through Gases*, adding that they can easily be corrected. Bohr makes his comments in a friendly, cheerful way in his halting English, but it soon becomes apparent that he's made a big mistake. He insists he wasn't trying to get the better of Thomson and that he's only here to learn. But it's too late. Thomson is miffed, and Bohr is disappointed at Thomson's lack of interest in being told his calculations are flawed.

Sometime later, Bohr asks Thomson to look over a paper of his. A few days later, he notices that Thomson has not even touched the manuscript, and asks him about it. This, too, is a breach of 'good manners'. Thomson makes it clear to Bohr that a young man of Bohr's age cannot possibly know as much about electrons as he does himself.

The stand-offish Englishman will no longer have any truck with Bohr, and even crosses the street to avoid him. No one wants to sit next to the sad-faced Dane during dinner at the long tables in the Trinity College dining hall. It's weeks before anyone even speaks to him again. 'Very interesting,' is the way Bohr later describes his time in Cambridge, 'but absolutely useless.' 'Very interesting, very interesting,' is Bohr's standard way of ending an exchange whenever he considers his conversation partner to be talking nonsense, speculating wildly, or defending dubious scientific hypotheses. 'Very interesting'—Bohr now wants to have no truck with Cambridge. At least he now has time to read the long novels of Charles Dickens, and his terrible English improves appreciably.

In February 1911, the Dane with the sad face has a real reason to be sad, when his father dies at the age of only fifty-four. Christian Bohr was a respected physiologist who studied gas exchange in the lungs during breathing. It was in his father's laboratory that Niels got his first taste of science. As Bohr mourns alone, he misses his fiancée, Margrethe Nørlund, the sister of an old friend. They met a year earlier at the Ekliptika debating club in Copenhagen, and decided shortly afterwards to marry as soon as they could. Niels writes Margrethe letters full of longing for her—at least one a day. In one such letter, he quotes a poem by Goethe:

> Spacious world and broadest life,
> Many years of honest quest,
> Ever probed and ever fathomed,
> Never closed, and often blest,
> Age preserved by loyalty,
> Fresh and friendly every day,
> Cheerful wits and pure intents:
> There! One comes well on his way.

One evening at a dinner, the lonely young Dane meets a forty-year-old man with a moustache, his greying hair parted to the side, and speaking in strangely accented English. Ernest Rutherford is the son of a Scottish farmer who emigrated to New Zealand. He's a professor at Manchester University, a winner of the Nobel Prize for Chemistry, and one of the world's top experimental physicists. He's powerfully built, with an equally powerful voice employed in swearing loudly whenever one of his experiments goes wrong. He's straightforward, which appeals to Bohr. Rutherford studied under J.J. Thomson, but now he's in competition with his former teacher. Who will be the first to fathom the construction of the atom?

Bohr realises that Manchester, rather than Cambridge, is the place for him. Rutherford has a greater understanding of the nature of atoms than Thomson, and his experiments are far more interesting. Rather than giving Bohr the British cold shoulder, Rutherford meets him with warmth and encouragement. 'He was almost like a second father to me,' Bohr will later say of Rutherford, even naming the fourth of his six sons Ernest in his honour.

In March 1912, Bohr finally manages to transfer from Cambridge to Manchester. He's determined to learn how to conduct experiments with radioactivity. However, Manchester is not destined to make a better experimental scientist out of him. He's not 'completely useless', but almost.

Radioactivity is the key to understanding the structure of the atom. Rutherford is convinced this is the case, just like his esteemed rival Marie Curie, with whom he otherwise loves to disagree. While working at the Cavendish Laboratory a few years earlier, Rutherford had discovered the explanation for alpha radiation. It consisted of particles that are far heavier than electrons, carrying the opposite charge and twice as much of it. When Rutherford and Hans Geiger captured these alpha particles and neutralised their charge, they realised the resulting particles were identical to helium atoms. Thus, when alpha decay occurs, a larger atom changes into a different, smaller atom by emitting a bit of matter that is very similar to the lightest form of the helium atom. But still nobody knew exactly what an atom is.

Rutherford realises that alpha particles can be fired like projectiles. They can be aimed at other things to find out what the target is made of. He and Geiger bombard a thin sheet of gold foil with alpha particles from a radioactive source. This makes it sound far more exciting than it is. The job of sitting for hours in the dark laboratory waiting for their eyes to get accustomed to the dark is left to Rutherford's co-workers. Their eyes need to be used to the dark in order to count the tiny flashes of light emitted whenever an alpha particle hits a phosphorescent screen.

The scientists are astonished by what they see. The majority of the alpha particles pass through the gold foil as if it weren't there. A few are deflected by a few degrees, like ricocheting bullets. But the finding that most surprises the physicists is that some alpha particles don't make it through the foil at all. They bounce back in the direction they came from. 'It was quite the most incredible event that has ever happened to me in my life. It was almost as incredible as if you fired a 15-inch shell at a piece of tissue paper and it came back and hit you,' Rutherford was to say later. Rutherford realises that those alpha particles must be bouncing off something that is even heavier than they are. This leads him to conclude that atoms have tiny, dense 'nuclei' where almost their entire mass is concentrated. The rest of the atom must be pretty empty. Rutherford famously describes the atomic nucleus as being 'like a fly in a cathedral'.

The idea seems to be right, but Rutherford still needs the right theoretical proof. J.J. Thomson still strongly believes in his 'plum pudding' model of the atom, which does not include the idea of a nucleus, but imagines the atom's electrons as being distributed evenly throughout a volume of positive charge, like the plums in that traditional English dessert. Rutherford does not have the means to change Thomson's mind. He's a laboratory physicist. He is as inept at juggling equations and formulae as Thomson is at conducting experiments.

One of Rutherford's assistants is called Charles Darwin. He's the grandson of the great evolutionary biologist and the only theoretician in Rutherford's team. Niels Bohr arrives in Manchester just as Darwin is trying to make theoretical sense of Rutherford's findings. Darwin suspects that the majority of the alpha particles get tangled and stuck in the jumble of electrons in the gold foil, losing their energy in the process. Only an exceptionally few alpha particles happen to collide with one of these theoretical nuclei and bounce back in the direction they came from. Darwin hopes this theory will help gain an understanding of the structure of the atom. He envisions electrons buzzing about randomly within the volume of the atom.

But the theory doesn't hold water. When Darwin tries to adapt his model to understand how alpha particles get tangled up in different materials, it generates nonsense. It predicts atoms of the wrong size. Seeing these nonsensical results, Bohr is reminded of his own doctoral thesis. He has a hunch that the reason the two theories don't stand up is the same: electrons don't buzz around as freely as both he and Darwin assumed. They are somehow attached to the nucleus of the atom. Bohr plays around with various images of the atom. He visualises electrons as tiny balls bouncing up and down on attached springs. Or as miniature planets orbiting the atomic nucleus like a sun. It's just a mental game, but Bohr is certain of one thing: electrons must be in motion. Otherwise, the atom would fall apart. But, on the other hand, if they are in motion, they must be emitting electromagnetic radiation, which means they must eventually run out of energy and come to a standstill. It's a paradox!

Then Bohr takes a bold theoretical step. To stabilise the atoms in his model, he determines that electrons can't just whizz around inside the atom with just any randomly changing amount of energy. Their energy changes only in fixed amounts (a limitation known as the quantisation condition): it changes only by a 'quantum' each time. The exact way he has arrived at this realisation remains his secret. Perhaps history has repeated itself, and Bohr has committed the same 'act of desperation' as Max Planck, albeit in a different scientific context. Bohr is familiar with the trick used by Planck to reach his formula for radiation eleven years earlier, and he's familiar with Einstein's light quanta. He candidly admits that he can't provide scientific justification for his suggested solution, just as Planck was unable to for his solution to the blackbody problem. The concept of 'quanta of energy' is still in the air, and it's still shrouded in an aura of mystery.

But it works: Bohr is now far better able to understand the 'braking manoeuvres' of alpha particles. He hastily drafts a research paper and then, after just three months in Manchester, he hurries home to wed Margrethe. On Thursday, 1 August 1912, the couple are married in Margrethe's hometown of Slagelse on the island of Zealand—not in the town's magnificent mediaeval church, but in the town hall. As an atheist, Bohr refuses to have a religious ceremony. The mayor of Slagelse happens to be on holiday at the time, and so Margrethe and Niels are married by the local police chief. The ceremony takes all of two minutes.

Bohr is pleased that he no longer has to write out his texts himself. He finds it difficult to write and think at the same time. He much prefers to speak. From now on, he dictates most of his research papers to Margrethe, who has a talent for languages and is able to correct his still-clumsy English. She had planned to become a French teacher, but now she's Bohr's secretary. Her work begins even before their honeymoon is over. The couple travel to Cambridge and Manchester, where Niels shows Margrethe his workplace. In Manchester, they are able to hand Ernest Rutherford the research paper in which Niels unravels the mystery of the atom. Rutherford is impressed.

The next few months show just how productive Bohr's hastily cobbled-together model of the atom is. It solves a puzzle that physicists have been trying in vain to crack for decades: the emission spectrum of hydrogen. Around a century earlier, scientists noticed that when sunlight is split into its rainbow of constituent colours by a prism, the spectrum is streaked with hundreds of black lines. They developed sophisticated equations to describe the patterns of those lines. By Bohr's time, spectral-line analysis has become a scientific discipline in itself—but no one has any idea what's causing the lines.

Bohr is now able to explain spectral lines almost in passing with his model of the atom. Just as gravity binds the planets to the Sun, electrical attraction keeps electrons in their orbits around the atomic nucleus. However, unlike the planets, electrons can jump from higher orbits to lower ones and vice versa, but only when the amount of energy they gain or lose in doing so obeys the quantisation condition. In Bohr's hands, the science of spectral lines becomes the science of electron jumps.

This explanation of the spectral lines of hydrogen is nice, but is it perhaps just a lucky shot? Bohr convinces his sceptics by making a spectacular prediction about helium, the second element on the periodic table. It is rare on Earth, but one of the main elements making up the Sun, as shown by lines in the Sun's emissions spectrum. The element's name comes from the Greek word for the Sun—*helios*. Helium atoms can lose one of their two electrons in the heat of our burning home star. The remaining electron can then jump between orbits, like those of hydrogen atoms. Bohr's model predicts that the frequencies in the helium spectrum will differ from those in the hydrogen spectrum by a factor of four. A British experimenter makes very precise measurements in the lab and, finding that they differ by a factor of 4.0016, concludes that Bohr's model must be wrong.

But Bohr is quick to respond. For the sake of simplicity, he has assumed the mass of electrons to be negligibly small compared to that of the atomic nucleus. When he includes the known values for mass in his equation, it generates a difference factor of 4.00163. This is a previously unmatched level of agreement between theory and experimental results, and it causes a sensation. News soon spreads that a great discovery has been made by a young Dane.

With this discovery, Niels Bohr founds the science of nuclear physics. His model provides solutions to long-unanswered questions, but it also raises new questions. How does an electron decide whether to jump between orbits, and which orbit to jump to? Once again, events in the quantum world seem to have spontaneously taken on a life of their own and, once again, they seem to override the principle of cause and effect. 'That business of causality plagues me a great deal, too,' Einstein would write to Max Born a few years later, when the mystery of what causes quantum leaps is still unsolved. Einstein is not alone in his concern. Physicists are delighted to play around with Bohr's model of the atom, but they secretly know that something is not right with it.

Bohr also recognises that his model cannot represent the whole truth. Probably not even half of it, he suspects. But his model is a clue on the path towards the truth. And, with his tendency to think like a detective, that's just what he is hoping for. He loves crime fiction, and is a voracious reader of murder mysteries. When they go travelling, the Bohrs often take along an entire suitcase full of nothing but crime novels. And so Niels Bohr knows that the most obvious suspect never turns out to be the killer.

THE NORTH ATLANTIC, 1912

The sinking of infallibility

On 10 April 1912, the *Titanic*, the giant steamship hailed round the world as unsinkable, sets sail from the port of Southampton on its maiden voyage, heading for New York. Thousands of miles away, the Caribbean is experiencing an unusually warm spring, which adds extra power to the Gulf Stream. At the same time, the cold waters of the Labrador Current carry hundreds of icebergs south from the Arctic Sea. A barrage of icebergs forms at the point where the two currents meet, and the course of the *Titanic* takes the liner right through it. The night of the fourteenth to the fifteenth of April is clear and starry in the North Atlantic. The metal alloy used to construct the hull of the ship becomes particularly brittle in the cold waters. The ship strikes an iceberg carried down from a glacier in western Greenland by the Labrador Current, tearing several holes in the hull below the waterline on the starboard side. Within a dramatic three hours, the *Titanic* sinks, taking with it the popular belief that science and technology are infallible. Of the 2,201 passengers and crew on board, only 711 survive.

The Italian physicist Guglielmo Marconi, the inventor of radiotelegraphy and a winner of the Nobel Prize, does not have to fight with his fellow passengers over places in the lifeboats. He turned down the offer of a free ticket for him and his family on the maiden voyage of the *Titanic*. He had a lot of work to do, and was keen to reach New York quickly, so had taken a different steamer across the Atlantic just three days earlier. A lucky decision.

Still, Marconi played a key role in this drama. He built the radio system on the *Titanic*, and both radio operators on board, Jack Phillips and Harold Bride, are employees of Marconi's company, not the shipping line. They send out the SOS call with the *Titanic*'s position to other ships, and even continue to send out distress calls after the captain releases them from their duties, giving up only when the telegraph room is flooded. Bride survives. Phillips is drowned. 'Those who have been saved, have been saved through one man, Mr. Marconi ... and his marvellous invention,' Britain's postmaster general was to say later. Electromagnetic wave theory saves lives. Who would dare to question it?

MUNICH, 1913

A painter moves to Munich

Just about the time Niels Bohr is presenting his first paper on his atomic model for publication in *The Philosophical Magazine*, a painter named Adolf Hitler and an unemployed merchant called Rudolf Häusler are on a train, fleeing Austria for Munich to escape military service. The two had lodged at the same men's hostel in the Austrian capital. On arriving in Munich, they roam the city looking for somewhere to stay. They spot a small sign in the window of a tailor's shop in Schleissheimerstrasse that reads 'small room for rent'. Hitler knocks on the door. It's opened by the tailor's wife, Anna Popp, who shows him the room on the third floor. Hitler accepts immediately. Hitler and Häusler move in with the Popps for a rent of three marks a week. Hitler paints a watercolour every day, sometimes even two: views of the city, which he hawks to tourists in the taverns every evening. He has no contact with Munich's lively arts scene, and never receives visitors. He spends the evenings reading inflammatory political pamphlets. When his landlady warns him to stay away from political books and to concentrate on painting, he answers, 'My dear Frau Popp, does one know what one needs and doesn't need in life?'

Meanwhile, Austria-Hungary launches a search for anyone avoiding conscription to the army. On 22 August 1913, the Vienna police publish a wanted notice for 'Adolf Hitler, recently residing in the Men's Hostel, Meldemannstrasse, current whereabouts unknown. Investigation ongoing.'

Just days before, on 17 August, the Emperor Franz Joseph had appointed the man due to succeed him, Archduke Franz Ferdinand, to the position of Inspector General of the Austro-Hungarian Armed Forces, considerably extending his power. The heir's great rival, General Chief of Staff Count Franz Conrad von Hötzendorf, is demanding a preventive war against Serbia and Montenegro. Franz Ferdinand refuses. Peace continues — for now.

MUNICH, 1914

On tour with the atom

In July 1914, Niels Bohr goes on tour again with his atom, but this time without his wife. His itinerary takes him to Göttingen and Munich. Göttingen is the bastion of pure mathematics and mathematical physics. Carl Friedrich Gauss, known as the *princeps mathematicorum*, the prince of mathematicians, lived and worked there in the nineteenth century. However, not much has moved on in Göttingen since Gauss's time, and the university's glory is starting to fade somewhat. Mired in tradition as it is, Göttingen is not an easy place for Bohr to sell his atomic model. As soon as rumours spread to Göttingen of Bohr's theory, the academic authorities there had rejected it as 'foolhardy' and 'abstruse'. But Bohr is now there in person to present his theory, with his soft voice, his deliberate way of speaking, and his awkward German. And he makes at least a little headway, since the reactions he receives are now not all completely negative. While Alfred Landé, assistant to the mathematician David Hilbert, calls Bohr's model 'nonsense', Max Born, who has recently received his professorship and who found Bohr's model incomprehensible when he saw it on paper, takes a milder tone after hearing Bohr's lecture, admitting, 'This Danish physicist looks so much like a real genius, there must be something to it.'

Bohr has an easier time of it in Munich. There, the field is ruled by Arnold Sommerfeld, a sixty-four-year-old professor of theoretical physics with the manner and the moustache of an officer of the Hussars. Despite spending some years in Göttingen, Sommerfeld has retained his inquiring mind and youthful curiosity. He was one of the first scientists to come out in support of Einstein's theory of relativity, at a time when other physicists of his generation were still refusing to rethink their conception of space and time. When he heard about Bohr's model, he wrote to the Danish scientist saying he was still sceptical, but had to admit that the predictive power of the model was 'unquestionably a great achievement'. Sommerfeld gives the Dane a warm welcome in Munich, and encourages his students to engage with Bohr's new atomic physics.

From 1912 to 1914, the German physicists James Franck and Gustav Hertz carry out research into the impact of an electron on an atom—known to history as the Franck-Hertz experiment. Using an electric field, they accelerate electrons in a glass vacuum tube, and fire them through a cloud of mercury vapour. They measure how much energy the electrons lose when they collide with the mercury atoms in the vapour. In May 1914, Albert Einstein becomes one of the first to realise that the measurements made by Franck and Hertz are a confirmation of quantum theory and thus support Bohr's model of the atom.

But one event at that time overshadows all the concerns of physics. On 28 June 1914, Serb nationalists assassinate the nephew of the emperor and heir presumptive to the Austro-Hungarian throne, Franz Ferdinand, along with his wife, Sophie. His uncle, Emperor Franz Joseph I, is not exactly crippled with grief. He considers Franz Ferdinand unfit for the position of emperor, which he believes is evident not least of all in his nephew's insistence on marrying for love. The emperor considered Sophie, a former lady-in-waiting at court, to be so far beneath Franz Ferdinand's station that he long resisted the marriage, and eventually allowed it to go ahead only under the condition that Sophie be known not as 'the future empress consort', as tradition dictated, but rather as 'the consort of the future emperor', and also under the condition that the couple's children bear Sophie's family name and have no claim to the throne. On his death, the headstrong Franz Ferdinand is succeeded as heir presumptive by the emperor's great-nephew Karl, who is far more traditionalist in his outlook.

The assassination of Archduke Franz Ferdinand does not go unpunished for long, and Austria even goes so far as to attack Serbia. This disturbs the delicate network of international treaties keeping

Europe in political equilibrium, and one country after another begins mobilising its forces. The 'July Crisis' escalates across the continent. The Central Powers around the German Empire and Austria-Hungary find themselves opposed by the Triple Entente, consisting of Great Britain, France, and Russia, who are soon joined by Italy.

In July 1914, Niels Bohr and his younger brother, Harald, travel from Munich to Tyrol to go hiking in the mountains. In the few newspapers they manage to get their hands on, they read increasingly worrying reports of an impending war. All travellers are hastily heading home, and the Bohr brothers are forced to accept that now is no time for mountain hikes. Just half an hour before Germany declares war on Russia, they cross the border back into Germany. They travel from Munich to Berlin on a train that is so overcrowded they have to spend the night standing in the corridor. On arriving in Berlin, they are hit by the full force of the people's support for the war. 'There was such boundless enthusiasm,' Niels Bohr later recalls, 'there was shouting and cheering over the fact that they were going to war again. Such enthusiasm is usual in Germany as soon as anything military is involved.' The brothers take a train to the Baltic Sea port of Warnemünde, and catch the last ferry to the safety of neutral Denmark.

The war having brought a sudden end to his entry into the world of German physics, Bohr is now back in Copenhagen, where he has no laboratory of his own and little time for research. He has to spend his time teaching physics to medical students, writing on the chalkboard, and writing for himself again, which is difficult for him. Margrethe can't take over this task for him in the lecture theatre.

With a war on, the Danish government has no time for Bohr's request that it set up a theoretical physics institute. So he gratefully accepts an offer from Rutherford to return to Manchester. It's almost like a homecoming for Bohr.

However, much has changed in the meantime, and nothing is as it was three years ago. Rutherford's laboratory seems deserted. Many of the researchers have gone off to war. James Chadwick happened to be in Berlin on a research fellowship when war broke out, and is now stuck there for the duration of the hostilities as a prisoner of war. Henry

Moseley, one of Rutherford's most gifted students, is shot and killed by a sniper at Gallipoli, during a failed invasion by the Entente powers and British Empire forces. Hans Geiger, who gave his name to the Geiger counter and who carried out the famous 'gold foil' experiments under Rutherford, is now an officer of the German artillery. As part of the 'gas troop', he's involved in preparing gas warfare. In France, meanwhile, Marie Curie and her daughter Irène are constructing mobile radiography units for the troops.

Rutherford, too, has little time now to think about atoms. He's developing a sonar system to detect the German submarines attacking British military and trade ships. Once again, Bohr is alone. War is not an ideal time for someone who is trying to understand the world.

Bohr finds this isolation particularly tough to deal with. His way of researching and of understanding the world has always been based on discussions with his colleagues, in a kind of continuous informal seminar. He relies on interaction with other scientists, on thinking aloud, throwing out ideas and then correcting them, digressing and stopping short to ponder. It's the start of a happy time with his new wife, but one that is lacking in scientific activity. Margrethe is happy in Manchester. The industrial city may not be as pretty as Cambridge, but she finds the people far friendlier there.

However, the war doesn't paralyse the scientific community completely. In Munich, now completely cut off from the world of international physics, Sommerfeld undertakes a thorough study of Bohr's atom. Essays and scientific journals circulate between enemy countries via obscure channels. Even when nations are shooting at each other, ideas can travel. Bohr is able to inspire other physicists, even across the trenches of war.

In Manchester, Bohr conceives the notion that some electrons might orbit the nucleus of their atom in elliptical paths, rather than all travelling in circular orbits. That could explain why the spectral lines of hydrogen sometimes split into a number of finer lines. But when he tries to do the calculations to support this idea, he comes up against a brick wall.

For a physicist of world standing, Bohr is a remarkably poor mathematician. Even a cursory glance at his research papers reveals how few mathematical equations they include. Instead, he begins with general concepts and assumptions, considers them in a rather philosophical way, and is very sparing with quantitative arguments and formal derivations. For most of his career, he relies on a range of mathematically gifted assistants to express his extraordinary physical insights as formal arguments. This way of working adds to the air of mystique that increasingly surrounds Bohr. He has a unique way of looking at things and an ability to identify immediately the critical questions and problems, and to envision a possible way to solve them. But he is often unable to develop those answers mathematically himself. Many years later, Werner Heisenberg would recall a conversation in which 'Bohr confirmed to me that he did not work out his complex atomic models with classical mechanics, but rather that they came to him intuitively as images, based on his experience'.

Thus Bohr is not able to develop fully his idea of elliptical orbits. He publishes only a very sketchy outline of his proposal, which somehow finds its way into the hands of the formally adept and ingenious Arnold Sommerfeld in Munich. Trained in the best German tradition, a master of mathematical methods and their application to the problems of mechanics and electromagnetism, Sommerfeld is the ideal man to take the next step.

If Bohr is the Copernicus of the atom, Sommerfeld is its Kepler. He calculates the intricate mechanics of this miniature, quantised planetary system, resulting in a convincing argument for the idea that electrons' elliptical orbits are constrained to certain values. The degree to which an orbit is flattened—that is, its ellipticity—is divided into quanta, as is its height. Using such tricks, Sommerfeld is able to further decipher those puzzling spectral emission lines.

Bohr is delighted to see that his atomic model is productive. 'I do not believe I have ever read anything with more joy than your beautiful work,' Bohr writes to Sommerfeld. Physicists begin to speak of the Bohr-Sommerfeld model, perhaps science's first truly global achievement. The core of the model's concept was provided by the New Zealander Rutherford; its configuration is the brainchild of Danish-born Bohr after he met Rutherford in England; and the fine details come from Sommerfeld of Germany. However, the model is also a bold mix of old and new physics, of classical and quantum mechanics; productive maybe, but probably wrong, as Bohr himself recognises.

Bohr returns to Copenhagen in 1916, this time for good. He's no longer the shy young student he used to be. The Danish state finally gives in to his constant lobbying, and the university where he once attended first-year undergraduate lectures provides him with his own professorship and even his own institute, although it initially consists only of one small office that Bohr has to share with his first assistant, the Dutch physicist Hendrik Kramers. In 1919, a third desk is installed in their little office, for their new secretary. But Bohr has bigger plans. After he successfully canvasses for the funding, construction begins on a new institute building in a small street in Copenhagen called Blegdamsvej. The large double doors of the three-storey building open for the first time in 1921, with 'UNIVERSITETS INSTITUT FOR TEORETISK FYSIK' emblazoned in large letters above the entrance. The building boasts an auditorium leading off to the right from the large entrance hall, as well as a library and a cafeteria.

Throughout the war years, many physicists wander the world in search of a peaceful, safe place to do science. Bohr finds that place in his native country. He is now one of very few professors of theoretical physics in the world, and soon becomes a celebrity in Denmark. In the 1920s, Bohr's institute is visited by more than sixty theoretical physicists, many of whom spend long periods of time, even years, there. They come from all over the world—from the USA, the Soviet Union, Japan. Most are young. Bohr personally arranges the funding for their visits. He establishes a new kind of cooperation that goes far beyond just physics. The scientists not only work together, but also live, eat, and play football together. Bohr takes them skiing, hiking, or to the cinema. He likes westerns best.

When the young Dutch physicist Hendrik Casimir moves to Copenhagen to study under Bohr, Casimir's father wants to test whether Bohr really is a national celebrity. So he sends a letter addressed simply to 'Casimir, c/o Niels Bohr, Denmark'. The letter arrives even before Casimir does, and the young scientist's father can rest assured that his son is in good hands.

Bohr never gets better at speaking in public. He stutters and stammers his way through his lectures, his mind leaping from one thought to another, 'Er, and, and, and … but, but …' For Niels Bohr, lectures are not a place for explaining old ideas, but for thinking aloud.

BERLIN, 1915

Good at theory, bad at relationships

Zurich, February 1914. Albert Einstein, now aged thirty-four, senses he's heading for a time of upheaval, both in his thinking and in his life. As he writes to his thirty-eight-year-old cousin and lover, Elsa Löwenthal, in Berlin: 'I haven't had time to write as I'm busy with really great matters. Day and night, I think more and more deeply about the things I have gradually discovered over the past two years, which are a tremendous advance in the fundamental problems of physics.' A few days later, he continues in another letter, 'My little boy has the whooping cough, an ear infection, and the flu, and is very poorly. The doctor insists he be taken south for a time, as soon as he's well enough to travel. There is a positive side to it, since Miza will have to go with him, and I will be alone in Berlin for a time.' Miza is Einstein's wife, Mileva. The Einsteins' marriage is not going well. 'I treat my wife like a servant who can't be sacked,' Albert tells his lover.

On 29 March 1914, a rainy Sunday, Einstein arrives in Berlin by train. There is nothing about him to indicate that he has come to the German capital to stay. He travels light, with only his violin case in his hand and his half-finished theory in his head as he walks along the platform. That theory will later become known as 'the general theory of relativity' and will turn Einstein into the most famous scientist in the world.

However, for now, that future fame is nothing but a promise. In scientific circles, Einstein is seen as the next Copernicus, but no one knows much more about him than that. Max Planck and Fritz Haber, the men who wield all the power at the Prussian Academy of Sciences, have long been working on persuading Einstein to move to Berlin. Unlike in Zurich, his time in Berlin would be entirely his own to dedicate to his research. He can teach at the university if he wants, but he doesn't have to. Einstein is looking forward to this life 'as an academy man, free of any kind of obligation, almost like a living mummy'. First, he has two weeks of freedom to spend with Elsa before Mileva joins him in Berlin with their two young sons.

One of Einstein's first meetings in Berlin is with the influential businessman and banker, Leopold Koppel. Koppel's foundation supports the newly established Kaiser Wilhelm Society for the Advancement of Science, as well as resisting the rise of social democracy and paying Einstein's salary. Koppel is also interested in financing a Kaiser Wilhelm Institute for Physics, with Einstein as its director. But that will take some time. For the time being, Haber settles Einstein into an office at the Kaiser Wilhelm Institute for Physical Chemistry and Electrochemistry in the quiet suburb of Dahlem. There, Einstein finds exactly what he needs: peace and quiet. After all, he's working on nothing less than the overthrow of Newtonian mechanics.

Einstein has been working on that coup for years. He realised in autumn 1907 'that all the laws of nature can be treated within the framework of the theory of special relativity—except the law of gravity.' For a long time, he failed to understand why that is, until, one day, the 'breakthrough' came as he was sitting at his desk in the Bern patent office, letting his thoughts run free. Einstein imagined a person falling off the roof of a house. During freefall, he doesn't feel the weight of his own body—he's weightless while still within the Earth's gravitational field. This thought experiment put him on the track that was to lead to his theory of gravity.

This sounds similar to the flash of inspiration experienced by the twenty-three-year-old Isaac Newton in 1666 when he saw an apple fall from a tree while sitting in his parents' garden, and suddenly realised that the force pulling the apple down to the ground must be the same force that keeps the Moon in its orbit around the Earth. Only, in

Einstein's vision, it is the observer himself who falls, not just the apple.

For years, Einstein tinkers with the mathematics, trying to find a way to express his thought experiment in equations. Almost nothing about his idea reaches the outside world — only a rumour that Einstein is onto something big.

Planck would prefer Einstein to support him in developing his quantum theory, or, even better, in finding a way to rid the world of the concept of quanta entirely. He advises Einstein not to continue to pursue his revolution in gravitational theory, as it will never come to anything.

'And even if it does come to something, no one will believe you,' Planck prophesises to Einstein. 'Why do we need a new theory of gravity, when we already have an old tried-and-tested one?' the conservative Planck asks.

Einstein knows why. Although Newton's laws of gravity describe the world observed so far, they contradict the principles of Einstein's theory of special relativity, which says that no effect can propagate faster than the speed of light. According to Newton's theory, the gravitational forces of two bodies work upon each other instantaneously, no matter how far apart they are. That would make them infinitely fast. Einstein wants to understand gravity as a field, like electricity or magnetism. He imagines the Earth producing a gravitational field, similar to an electric or magnetic field, and that the field can affect apples, people, and planets alike. The only difference from an electromagnetic field would be that gravity always attracts and never repels, which complicates matters considerably.

The key idea had already come to Einstein in 1907, while he was working at the patent office in Bern. In freefall, the inertia of the falling body precisely compensates for its acceleration due to gravity. Conversely, acceleration — in a car, for example — feels exactly like the force of gravity. Inertial mass is the same as gravitational mass: Einstein has been developing this 'equivalence principle' into a full theory for years, and he has made much progress by the time he alights on the railway platform in Berlin.

Einstein has calculated that rays of light travelling through space are more heavily bent as they pass by celestial bodies than Newton's theory predicts. Thus, precisely measuring such light deflection could provide proof of whether Einstein's theory is more accurate than Newton's. Einstein attempts to find this proof in existing photographic plates of solar eclipses. When the sun's light is blocked by the moon, the stars right next to the sun become visible. But the quality of the photographic plates is too poor.

A more promising chance of success is offered by the total solar eclipse that will be visible in northern Canada, northern Europe, and parts of Asia on 14 August 1914. Astronomer Erwin Freundlich, one of the few scientists who has taken Einstein's theory seriously, sets out for Russia in April of that year, taking all manner of scientific equipment with him. But when war breaks out, instead of photographing the solar eclipse, Freundlich finds himself taken prisoner by the Russians.

Meanwhile, Einstein's marriage is falling apart. Albert's attitude to Mileva is increasingly chauvinistic. 'You will see to it that I am served three regular meals a day IN MY ROOM,' he instructs her in writing. 'You will renounce all personal relations with me. You will stop talking to me immediately if I request it.' He uses the very authoritarian language with his wife that he so detests in Prussians. Albert and Mileva separate at the end of July 1914. She returns to Zurich with their sons, while he moves into a small apartment in the Wilmersdorf district of Berlin. Albert writes to his beloved Elsa, 'I can't come and see you. We must appear to be saintly at this time.' His apartment is only fifteen minutes' walk from Elsa's.

Ironically, Einstein has arrived in Berlin at a time when the German capital is raging with nationalism and enthusiasm for war; ironic, since this is the man who renounced his German citizenship due to his abhorrence of all things militaristic, and who lived for years in exile in Switzerland, first as a stateless person, then as a Swiss citizen, with a period of dual Austrian citizenship in between. Now he's back in his homeland, which has become foreign to him. He's a Prussian civil servant.

'Europe in its madness has now embarked on something incredibly preposterous,' Einstein writes to his friend Paul Ehrenfest. 'At such times one sees to what deplorable breed of brutes we belong. I am musing serenely along in my peaceful meditations and feel only a mixture of pity and disgust.' He advocates for the establishment of a League of Nations, and joins the New Fatherland Confederation, a German pacifist association advocating for peace through diplomacy and democratic reform.

However, Einstein is rather alone in his anti-war stance. Many of his colleagues are swept up in the nationalistic euphoria. Even his friend the Austrian physicist Lise Meitner is surprised, writing after spending an evening with him in 1916: 'Einstein played the violin and spouted such delightfully naïve and peculiar political and belligerent views.' Max Planck exhorts his students to enter the trenches to fight against 'the breeding grounds of creeping insidiousness'. Walther Nernst, now aged fifty, signs up as a volunteer ambulance driver. Fritz Haber puts his formidable abilities as a chemist to use developing weapons, and becomes one of the lead scientists preparing for gas warfare. While Haber is on the Western Front, preparing the first large-scale gas attack, Einstein gives Haber's twelve-year-old son extra lessons in mathematics. The gas attack involves releasing 150 metric tons of chlorine gas on the defenceless soldiers manning the French positions. Fifteen hundred troops are killed. Shortly after the attack, Haber's wife shoots herself with his army pistol, but an undeterred Haber continues to develop gas weapons, turning his entire institute, with its staff of 1,500, over to this purpose.

Planck, Nernst, Röntgen, and Wien are among the ninety-three signatories of an appeal 'To the Civilised World!' published in the major German newspapers and further afield on 14 October 1914. In it, the signatories protest 'against the lies and calumnies with which our enemies are endeavouring to stain the honour of Germany in her hard struggle for existence—a struggle that has been forced on her'. They deny that Germany is to blame for starting the war and that it breached Belgium's neutrality—something even the Kaiser has admitted. 'Have faith in us! Believe that we shall carry on this war to the end as a civilized nation, to whom the legacy of a Goethe, a Beethoven, and a Kant is just as sacred as its own hearths and homes.'

Einstein had hoped that his sponsor Planck would not join such voices. 'Even the scholars of various countries are acting as if they had their brains amputated eight months ago,' he lamented in a letter.

As Einstein's colleagues see it, he is fighting a losing battle on two fronts: against the war, and for a new theory of gravitation. But he now has a formidable competitor: David Hilbert, the legendary Göttingen-based mathematician, is also working on a new theory of gravity. 'Physics is far too hard for physicists,' says Hilbert. This is a provocation for a physicist like Einstein.

Einstein is initially flustered by this, but he soon enters into a creative frenzy. He makes discreet enquiries among friends as to Hilbert's progress. Hilbert may be the better mathematician, but Einstein has a formidable physical imagination. In his mind, he sees gravitational fields as emerging from curvatures in the structure of spacetime, which he described as 'a sheet floating (at rest) in the air'. Celestial bodies bend spacetime in the same way balls thrown onto a suspended cloth would dent it. Expressing these ideas in formal mathematical terms takes even Einstein to the very limits of his mathematical abilities. The genius does not make progress with flashes of inspiration, but with hard work, dedication, and perseverance. Bit by bit, he pieces his field equations together, then presents a version of his theory at a session of the Academy, which he immodestly extols: 'Scarcely anyone will be able to resist the magic of this theory, once he has understood it.'

At the very next session, just one week later, Einstein takes the academicians by surprise once again by presenting the extended version of his theory. Another week goes by, and Einstein brings a spectacular discovery to the session. His theory can explain the mysteriously convoluted orbit of the Sun's closest planet, Mercury, called the planet's perihelion procession. This is a pointer that Einstein is on the right track. Another week goes by, and, by the time of the session on 25 November 1951, Einstein has expanded his field equations once more—for the final time. The theory is complete, after eight years of constantly wrestling and wrangling with it. 'This means the General Theory of Relativity is now a completed logical structure,' he says. He makes no mention of Hilbert's contribution.

David Hilbert had handed in his calculations for publication on 20 November 1915, five days before Einstein, but his work actually appears later. The race is a draw. Einstein seeks a reconciliation with Hilbert: 'There has been a certain ill-feeling between us, the cause of which I do not want to analyse. I think of you again with unmarred friendliness and ask you to try to do the same with me. Objectively it is a shame when two real fellows who have extricated themselves somewhat from this shabby world do not afford each other mutual pleasure.' Hilbert accepts the offer of friendship, but continues to insist for the rest of his life that he was the one who first came up with the formulae for gravitation.

GERMANY, 1916

War and peace

It is 1916; the war is in its third year. An invisible rift runs through Germany, splitting families, dividing friends. The physicists Friedrich Paschen and Otto Lummer had once worked amicably together in the optics laboratory of the Imperial Institute for Physics and Technology. But that's over. Lummer is full of enthusiasm for the war; Paschen is for peace.

The most important physicists swept up by this nationalist zealotry include Wilhelm Wien, Johannes Stark, and Philipp Lenard. Wien writes an appeal to 'fight English influence in physics'. In August 1914, Lenard publishes a pamphlet on 'England and Germany at the time of the Great War', in which he writes, 'Were we able to destroy England completely, I would not consider it a sin against civilisation. Thus, away with any consideration for England's so-called culture. Away with any timidity before the graves of Shakespeare, Newton, Faraday!' And Lenard is not alone. His words express what many German professors are thinking.

But the German-born professor Albert Einstein thinks differently. He writes an article in a 'patriotic remembrance book' published by the Berlin Goethe Federation with the title *The Country of Goethe, 1914–1916*. Contributors to this weighty volume include Paul von Hindenburg, Walther Rathenau, Ricarda Huch, Sigmund Freund, and others. Many are full of patriotic bluster and warlike clamour. One of the few to express cautious concern is the dramatist Elsa Bernstein, writing under her pseudonym Ernst Rosmer, 'God created death, man created murder.' Einstein is more direct:

> One can ponder the question: Why does man in peacetime (when the social community suppresses almost every representation of rowdyism) not lose the capabilities and motivations which enable him to commit mass murder in war? The reason seems to me as follows. When I look into the mind of the well-intentioned average citizen, I see a moderately lit, cosy room. In one corner stands a well-tended shrine, the pride of the man of the house, and every visitor is loudly alerted to the presence of this shrine upon which is written in huge letters 'patriotism'. However, it is usually taboo to open this shrine. Moreover, the master of the house barely knows, or does not know at all, that this shrine holds the moral requisites of bestial hatred and mass murder, which he dutifully takes out in case of war in order to use them. This type of shrine, dear reader, you will not find in my little room, and I would be happy if you would adopt the attitude that in that same corner of your little room a piano or a bookshelf would be a more fitting piece of furniture than the one you find only tolerable because you have been used to it since your early youth.

Other physicists open up that shrine of patriotism with gusto. In 1919, Philipp Lenard hears the speeches of the first leader of the Nazi Party, Anton Drexler, as well as those of Adolf Hitler. In February 1920, he visits 'the Party's first mass event', and is impassioned by it. In 1926, Lenard travels to Heilbronn to attend a party convention and to meet Hitler personally. And in 1928, Hitler pays a visit to Lenard in his Heidelberg home; for Lenard, it is to become one of the most memorable events in his life.

BERLIN, 1917

Einstein breaks down

The hard intellectual labour, his bachelor's lifestyle, and his sadness over the war take their toll on Albert Einstein's health. In February 1917, he collapses with severe stomach pains. He's diagnosed with liver failure. His condition continues to worsen over the next two months, and he loses twenty-five kilograms in weight. It's the start of a series of ailments—including jaundice, gallstones, and a life-threatening duodenal ulcer at the exit to his stomach—that will afflict him in the coming years. At not even thirty-eight years of age, he becomes seriously concerned about his 'decrepit body'. His doctor orders him to 'take mineral waters and keep a strict diet'. That's easier said than done. Prussia is starving. After crop failures and the 'Turnip Winter' of 1916–17, even potatoes are in short supply. There's substitute bread made from blood and sawdust, substitute jam made from turnips, substitute butter made out of suet, coffee made from chestnuts, spices made from ash, and substitute everything else: soap made out of clay, clothing made from paper—Germany has become the substitute nation. The authorities recommend roast crow as a replacement for chicken. Cats, rats, and horses all end up in the sparely heated ovens of Germany. Eighty-eight thousand people die from starvation in the country in 1915. The next year, that number reaches 120,000.

However, Einstein is lucky compared to many other people in Berlin. His relatives in Switzerland send him food parcels. Elsa nurses him. He spends the summer of 1918 convalescing in the fishing village of Ahrenshoop on the Baltic coast. While there, he doesn't work, doesn't take any calls, doesn't read any newspapers. Instead, he sunbathes, and takes short, leisurely, barefoot walks on the beach. His painful attacks abate, and by the winter semester of 1918–19 he's able to resume his work at the university. He gives a course of Saturday-morning lectures on the theory of relativity, but he soon has to cancel it 'due to the revolution'. On 3 October 1918, the German imperial government petitions the American president, Woodrow Wilson, for peace. Wilson's response is to demand that Germany become a democratic country.

Peace and democracy: both good news for Einstein. But a shock for many Germans, who have believed to the very end that they would emerge victorious from the war. The army is decimated and exhausted. On 4 November, sailors revolt in the port city of Kiel. Their uprising grows into a revolution and spreads throughout Germany, reaching Berlin on 9 November. Workers' and soldiers' councils are formed, and a general strike is called. Protesters outside the Reichstag demand an immediate end to the war. On the Saturday when Einstein's lecture is cancelled, a republic is declared. The following night, the Kaiser abdicates and flees to the Netherlands. Einstein sends euphoric postcards to his relatives in Switzerland: 'The Great Thing has happened! … We have thoroughly rid ourselves of militarism and sleepy officialdom.'

Together with Max Born and the psychologist Max Wertheimer, Albert Einstein takes the tram to the Reichstag, the manuscript of a speech to the 'comrades' in his pocket. The armed revolutionaries in front of the Reichstag recognise him and let him through to the freshly proclaimed German president, Friedrich Ebert. Einstein manages to obtain the release of the rector of the university, who has been imprisoned by the revolutionary student council. He does not manage to make his speech.

BERLIN, 1918

Pandemic

Haskell County in the US state of Kansas is a sparsely populated area. Sandstorms often sweep across the bleak plains. This is mainly farming country, and most of the farmers keep chickens. In February 1918, a doctor called Loring Miner finds himself making an unusually large number of house calls. He rushes from farm to farm, treating people with sudden-onset severe flu symptoms: coughing, fever, a rattling in the lungs. What most surprises Dr Miner is that many of these patients are young and were previously in robust health. This is not the usual seasonal flu. Miner sends a warning to the Public Health Service, but receives no reply.

On the other side of the Atlantic, in March of the same year, just as the German army is preparing to launch its spring offensive against France, the flu epidemic reaches Camp Funston at Fort Riley. An army cook falls ill: his job includes dishing out food to recruits at the base as they prepare to join the war in Europe. The recruits are then transported across the Atlantic in cramped ships, and end up living in miserable, unhygienic conditions in the trenches, where it's muddy, filthy, cold, wet, and infested with rats and other vermin. The young men's immune systems are unprepared to cope with the mutated influenza virus that biologists will later allocate to the H1N1 type. There are several thousand new infections among the soldiers every day, with a mortality rate of 10 per cent. Many of the infected men are taken prisoner by the Germans, and thus the virus is transported across the front lines. In no time, it puts 900,000 German troops out of action. There is little that the military doctors on either side can do.

The first mention of the virus in the newspapers appears in late spring, on 27 May 1918. The Spanish press reports that 'a strange illness with an epidemic character' has appeared in Madrid, and has infected the Spanish king, Alfonso XIII. So the virus becomes known as 'the Spanish flu', although it has not originated in Spain at all. As a neutral country in the First World War, Spain is not subject to military censorship and war propaganda, so, unlike those in the warring nations, Spain's newspapers are able to report freely about the epidemic.

But the virus is initially known by many names as it spreads around the globe. The British call it 'Flanders fever', since that's where British soldiers first got infected. The Poles call it the 'Bolshevik disease', while among Germans it's known as 'blitz catarrh' due to its rapid progression. In Spain, they speak of the 'Portuguese disease', while in Senegal it gets the name 'Brazilian illness'. The *New York Times* writes of a 'German flu', since it seems to infect mainly Germans. Every country has its own idea about who is to blame.

Soldiers on home leave transport the virus into the German Empire, where it encounters a starving and demoralised population. The first wave is relatively mild, and most of those who get infected survive the illness. But when the second wave hits the empire — soon to be a republic — in the autumn, it hits hard. Within just a few months, hundreds of thousands of people are getting infected every day. Four hundred thusand people die, with 50,000 fatalities in Berlin alone. Around the world, 50 million people die of this flu virus — twice as many as lost their lives in the First World War.

No one is safe. In Munich, the virus kills the philosopher and economist Max Weber; in Vienna, the painter Egon Schiele also succumbs to the Spanish flu. And in Prague, the writer Franz Kafka, whose lungs are already ravaged by tuberculosis, is confined to his bed for weeks by it — but survives.

Flu, famine, and a gruelling war on two fronts: it's all too much for Germany. On 11 November 1918, French and German delegates sign an armistice agreement in the forest of Compiègne, ninety kilometres north-east of Paris. Suddenly, the battlefields fall silent after four years of artillery thunder. Germany's capitulation comes as a shock to many of its citizens. They hadn't even known that the *Reichswehr* was on the defensive. Weren't they reporting territorial gains on the Western Front only this summer? General Field Marshal Paul von Hindenburg utters the fateful statement, 'The German army has been stabbed in the back.' This back-stabbing myth, claiming that Germany lost the war due to the treachery of its own war-weary population, is to become one of the factors that, within twenty years, will lead Germany to start another world war.

In November 1918, communists take over Bavaria in a revolution. Kurt Eisner takes power, deposes the king, and proclaims the Free State of Bavaria. Rosa Luxemburg and Karl Liebknecht proclaim a socialist republic in Berlin. Luxemburg, Liebknecht, and Eisner are all murdered in early 1919. The attempted revolutions are followed by a counter-revolution: an attempted right-wing coup led by Wolfgang Kapp in March 1920, which fails. The industrial Ruhr Valley is rocked by worker uprisings, which begin with the aim of warding off the right-wing coup, but soon turn into another grab for power. Children on the streets throw stones and even fire guns. The military, aided by volunteer paramilitaries, stamps out the insurrection with brute force. Executions and mass shootings are carried out under martial law. Anyone who is found to be carrying a weapon when arrested faces the firing squad, even if they're wounded.

The moderate German Jewish political leader Walther Rathenau is assassinated in June 1922. The authorities in the newly formed Republic of Baden order universities to close as part of a period of national mourning. But in Heidelberg, Nobel Prize–winning physicist (and later Nazi supporter) Philipp Lenard refuses to fly the flag at half-mast, or to cancel his seminar, clearly demonstrating his opinion that a dead Jew is not worth giving his students the day off for. Although disciplinary proceedings are opened against Lenard, they are soon dropped.

The German police have difficulty maintaining order, since the Allies imposed limits on the number of German military and police personnel in the Treaty of Versailles. The Allies fear armed policemen can easily become soldiers.

All around the world, people have lost dear friends and family members—to war, famine, or flu. For some, these experiences have shaken their faith in science and technology as the path to progress. The result is a boom in spiritualism and superstition. Widows, orphans, and bereaved parents long for contact with the dead. Two of the most prominent exponents of spiritualism in Britain are Sir Arthur Conan Doyle, creator of the master detective Sherlock Holmes, and Sir Oliver Lodge, a physicist working on radio waves. Lodge's son was struck by shrapnel and killed in Belgium. Doyle's son was seriously wounded in France and died of pneumonia. After the end of the war and the pandemic, the two men tour Britain and America, showing people how to make contact with the hereafter and speak to their dead loved ones. Doyle is quoted as saying that hearing from his son during a seance in 1919 is 'the supreme moment of my spiritual experience'.

The mighty scientists are humbled. 'Science has failed to protect us,' the *New York Times* avers. There can be no knowledge without uncertainty—and that will become the core concept in the next great theory of physics.

THE MID-ATLANTIC, 1919

The moon obscures the sun

On 14 February 1919, the marriage of Albert and Mileva Einstein is dissolved before the district court of Zurich, on the grounds of 'incompatibility of natures', after five years in which Albert pressed for the divorce and Mileva resisted, until Albert offered to pay her more maintenance. He promises her the entire purse of the Nobel Prize. He hasn't actually won it yet, but Einstein is convinced it's only a matter of time. On 2 June, he marries his cousin Elsa Löwenthal. He is forty; she is three years older. But by this time he has also begun an affair with Elsa's daughter Ilse, who calls Einstein 'father Albert'. It's just the latest in a long line of Einstein's affairs — and it won't be the last.

Little does Elsa suspect that the events of the coming months will change the lives of the newlyweds completely, and catapult Einstein to world fame.

In February 1919, shortly after Albert and Mileva's divorce is finalised, the British Royal Society sends out two expeditions — one to the village of Sobral in northern Brazil, and the other to the island of Principe off the coast of Spanish Guinea in Africa. The aim of both expeditions is to observe the total solar eclipse due on 29 May. Astronomers have calculated that these two locations should offer a particularly good view of the event. On the morning of 29 May, the leader of the project, Arthur Eddington, and his team are in a coconut plantation on Principe. It's raining heavily, and the clouds only begin to part around noon, when the eclipse is already underway.

The scholars manage to take two serviceable photos of the event, while their counterparts in Sobral snap eight images. Once back in England, Eddington uses the photographic plates to measure how much the moon-obscured sun bends the light of stars as it passes by. His results agree precisely with those predicted by Einstein's gravitational theory.

Einstein becomes a global star overnight. The president of the Royal Society, J.J. Thomson, tells a British newspaper that the theory of relativity 'opens up a whole new continent of scientific ideas'. Einstein is also celebrated in post-war Germany, with articles about him and his theory of relativity appearing everywhere. The weekly news magazine *Berliner Illustrirte Zeitung* puts him on a par with Copernicus, Kepler, and Newton. The *Times* of London runs the headline 'Revolution in Science, New Theory of the Universe, Newtonian Ideas Overthrown'. A variety theatre, the London Palladium, invites Einstein for a three-week guest performance, but he declines the offer. One young woman faints at the sight of Einstein. 'Everything is relative' becomes a buzzword for popular culture, the Roaring Twenties, and Americanisation.

But there are critical voices mixed in among all the exaltation. Some are friendly; others are more malicious. J.J. Thomson tells a journalist that 'no one has yet succeeded in stating in clear language what the theory of Einstein's really is'.

A deep sense of insecurity in Europe created during the First World War by the carnage, deceptive propaganda, social hardship, and disappearance of traditional ways of life all seems to be encapsulated in the theory of relativity. A counter-movement arises, based on nationalism and calling itself '*Deutsche Physik*'. Its proponents, foremost among them the Nobel laureate Philipp Lenard, reject modern theoretical physics as 'Jewish', and dream of a purely 'Aryan' science. As a Jew, a theoretician, and a pacifist, Einstein is the embodiment of everything they despise.

Lectures given by Einstein in Germany frequently end in commotion. When he makes admission to his lectures free for Jewish refugees from Eastern Europe, anti-Semitic students go on the rampage. 'I'll slit that filthy Jew's throat,' one of them is heard shouting. Einstein begins to receive letters containing death threats.

But he remains undaunted. Never having thought much about his Jewish heritage before, Einstein is now forced to become more conscious of it by the anti-Semitic hostility he's now receiving. He will first come into contact with Zionist organisations in the 1920s, although he doesn't share all their aims—he was never concerned with the foundation of a Jewish nation state. But he supports cultural Zionism. He believes that Palestine should become a safe haven for persecuted Jews, in peaceful coexistence with the Arab population, and that it would thus become a symbol, encouraging more self-confidence among people of the Jewish diaspora. Einstein supports the establishment of the Hebrew University of Jerusalem, which opens officially in 1925. In 1929, he will come out in support of abortion rights and the decriminalisation of homosexuality, and against 'mystery-mongering in sexual education'. If only he would treat the women in his life better.

MUNICH, 1919

A young man reads Plato

Munich, in the spring of 1919. While Arthur Eddington is on the island of Principe in the Gulf of Guinea, waiting for the eclipse to happen, First World War veteran Franz von Epp is leading his troop of right-wing paramilitary volunteers known as the *Freikorps* as they advance from Württemberg to Munich to smash the newly formed Bavarian Soviet Republic. The republic had been proclaimed after the assassination of the socialist prime minister Kurt Eisner by an aristocrat while Eisner was on his way to announce his resignation.

It's not only Munich that's in turmoil, but the whole of Germany. In January 1919, after elections are held, the National Assembly retires from Berlin to the city of Weimar. The delegates want to avoid the unrest in the capital, and are seeking a safe and peaceful place where they can draft a new constitution. They choose the former home of Johann Wolfgang von Goethe, Weimar, which will give its name to the republic founded there.

The fledgling republic attempts to impose order in Bavaria. Government troops surround Munich. As the city resounds with the salvos of gunfire, and smoke from the burning barricades swathes its streets, a seventeen-year-old schoolboy lies on the roof of a house in Munich's Bohemian district of Schwabing in the spring sun, reading Plato's *Timaeus*, the dialogue in which Socrates claims that all the world is mathematics. The name of this youth is Werner Heisenberg.

Werner Karl Heisenberg was born in the Franconian university town of Würzburg in 1901, a year and a half after his brother, Erwin. Their father, August, came from a family of craftsmen in Westphalia, who once bore the name 'Heissenberg', but are now officially called 'Heisenberg' due to a spelling error on the part of a registry clerk. August has worked his way up to become a teacher of classical languages at the Old Grammar School in Würzburg. However, his sights are set even higher: his ambition is to become a university academic. As a loyal citizen of Bismarck's Germany, he is the embodiment of Protestant values: diligence, discipline, thrift, self-control, rationality, reading, and musical appreciation. The Heisenbergs go to church on Sundays, not because they are a religious family, but because they know it is their duty to do so. The father's stoic façade belies a temperamental character, of which Heisenberg's family often bears the brunt. He drives a wedge between his two sons, who will see each other as rivals for the rest of their lives. Erwin beats Werner in every subject at school and in every type of sport. The only exception is maths.

In 1910, August Heisenberg accepts an offer from the university in Munich. Werner attends the famous Maximiliansgymnasium in Ludwigstrasse, of which his grandfather had been school director just a few years earlier. Escaping the clutches of this family is not an easy thing to do.

Werner excels in maths, and attends university lectures even while still a high school student. Indeed, he occasionally stands in for the mathematics master at school. As the First World War draws to an end, Werner is preparing for his school leaving exams. While perusing a physics book, his attention is caught by a diagram depicting atoms as little balls of matter that attach to and detach from each other with hooks and eyes. Surely that can't be correct? The image must be a creation of the illustrator's imagination, rather than based on any scientific knowledge. If atoms are the smallest building blocks of matter, what are the hooks made of? Could Plato be right? Is it mathematics that holds the world together? On the other hand, mightn't that idea also be nothing more than an image with no basis in actual experience?

The war and its conclusion initially curb Heisenberg's ambitions. Life now is about pure survival, and science must wait. Heisenberg works as a farmhand in southern Bavaria. After the Bavarian Soviet Republic is quashed, he volunteers for auxiliary military service as a scout with the cavalry riflemen, and joins the newly founded, nationalist New Scouting movement, which aims to find healing in nature after the ravages of the war. It will later be consumed by the Nazi youth organisations. During long hikes around Lake Starnberg and through the Franconian countryside, he and his friends discuss atoms, geometry, and Einstein's theory of relativity.

Even at the age of eighteen, Heisenberg knows what he wants to do with his life: become a mathematician. That's the science that offers him a way of understanding the world. Having wasted enough time already, he's anxious to progress quickly and skip the foundation courses of his university degree. For him, that's just kids' stuff that he mastered long ago while still at school.

Following a request from August Heisenberg, the Munich-based mathematician Ferdinand von Lindemann agrees to a meeting with Werner. Lindemann is known to history as the mathematician who proved that squaring the circle is geometrically impossible. He's a cantankerous old man with a white beard and outmoded views. He believes mathematics has a monopoly on the privilege of beauty, and that anyone who is serious about studying it must be instilled with this eternal truth.

Their conversation takes a bad turn even before it has really begun. As the shy young Heisenberg enters Lindemann's gloomy, antiquated study, he doesn't immediately notice the small, black dog sitting on the desk, eyeing him malevolently. On noticing the dog, Heisenberg is startled. The dog is also startled. Heisenberg falteringly explains that he's here to request that Professor Lindemann let him attend one of his seminars. The little dog starts barking at him, and even Lindemann is unable to quieten it. Lindemann asks Heisenberg what books he has been studying lately. The dog continues its yapping. Heisenberg tells the professor of his excitement at reading the book *Space, Time, Matter* by the mathematician and physicist Hermann Weyl, on the general theory of relativity. A physics book! 'Then you are already ruined for mathematics,' says Lindemann, and promptly ends their interview. An eighteen-year-old stripling with the audacity to debase mathematics by applying it to the perceivable world is, in his view, unworthy of his support. Heisenberg leaves the room to the continued sound of the little dog's yapping.

That is the end of Heisenberg's affair with mathematics. Disappointed, he consults his father, who suggests he try studying mathematical physics rather than pure maths. Once again, Heisenberg's father uses his connections, and arranges for his son to visit Arnold Sommerfeld, who receives Heisenberg in his bright and airy office. Sommerfeld has the severe look of a Prussian army officer: small and wiry, but with a pot belly, and a moustache waxed and twirled in the military style. He has just published a book with the title *Atomic Structure and Spectral Lines*, which is soon to become like a bible in the nascent field of atomic physics. For his students, Sommerfeld's enthusiasm for this field is infectious. There is no yapping dog in Sommerfeld's office, but, for Heisenberg, there is everything he missed at the interview with Lindemann: kindness and respect.

BERLIN, 1920

Great minds meet

Berlin, 27 April 1920. An unpleasant mix of excitement and expectation fills Niels Bohr's stomach as he makes his way from the railway station to the university. There's a melancholy aspect to the streets of the German capital, and the horses harnessed before the delivery carts look half-starved. Every now and then, an automobile jolts over the cobblestones, belching exhaust fumes. War-wounded veterans hobble aimlessly through the city: some on crutches, others with an empty sleeve dangling at the side of their overcoat. Women and children peddle cigarettes, matches, and socks to passers-by. Post-war deprivation has turned the people of Germany into a nation of traders. The smell of destitution is in the air; starving people generally don't wash. It's a wonder that any physics is even being pursued in Germany right now. Scientific journals are a rarity; books, an unaffordable luxury for many researchers.

And then Bohr is in the presence of the two men who've been awaiting his arrival: Albert Einstein and Max Planck. The two could hardly be more different in appearance, but each is friendly in his own way. Planck is the embodiment of Prussian formality and propriety, while next to him stands Einstein, with his wide eyes and wild hair, and trousers that are just a little too short at the ankle.

The war years have been tragic and painful for Planck. His son was killed in action, and both his twin daughters died soon after giving birth. Science has become his surrogate family. Together with colleagues, he founded the Emergency Association of German Science, which received funding from the government, from industry, and from abroad, and distributed the finances among scientists. 'As long as German science is able to continue to progress as it has until now, it is unthinkable that Germany will not remain among the group of civilised nations,' writes Planck in a 1919 newspaper article.

Planck invites Bohr to stay with him for the duration of his Berlin visit, and Bohr accepts. Soon, the three men's conversation turns to the topic that connects them all: physics. The anxious feeling in Bohr's stomach begins to dissipate.

Unlike many European physicists, Bohr, as a citizen of a neutral country, feels no post-war resentment towards his fellow physicists in Germany. Rather, he wants to do all he can to re-establish the ties that were cut by the hostilities. While German physicists are still barred from attending international scientific gatherings, Bohr takes the initiative and invites Arnold Sommerfeld to Copenhagen. It is shortly after this that Bohr receives his invitation from Max Planck to visit Berlin.

Visiting Einstein's home in Haberlandstrasse, Bohr brings butter and other food as a gift for his hosts. Not substitute butter. The real stuff! Einstein thanks him for the 'splendid gift from Neutralia, which still drips with milk and honey'. Elsa Einstein 'rejoices at the very sight of such delicacies'.

Then they get down to business, discussing radiation, quanta, electrons, and atoms. Bohr and Einstein fail to find any answers, but they do agree over which puzzles still need to be solved. With knowledge in the field in such a jumbled state, little more could be hoped for from this initial encounter.

Bohr spends his time in Berlin doing what he loves best: talking about physics from morning till night. It's a welcome change from his duties as head of the institute in Copenhagen. He particularly loves the lunch that young physicists at the university arrange for him. James Franck, Gustav Hertz, and Lise Meitner organise it as a picnic, since they know how much Bohr loves the outdoors. The chemist and Nobel Prize winner Fritz Haber offers to host the picnic at his country house. Max Planck provides the food, which the young scientists could never afford themselves. Meitner and the others had wanted the picnic to be 'bigwig-free' so they could have Bohr all to themselves. It's an opportunity to question him about his lecture to the German Physical Society, which left them rather depressed and feeling that they barely understood any of it. But here are Haber and Einstein sitting on the picnic blanket, and once again it's the 'bigwigs' who dominate the discussion with Bohr over waves and particles.

Einstein does understand what Bohr is saying, but he doesn't agree with it. Like almost everybody else, Bohr doesn't believe that Einstein's light quanta actually exist. Like Planck, he has accepted that radiation is emitted and absorbed in packets. But radiation itself is not quantised, Bohr and Planck insist. The evidence showing that light is a wave is too strong for them to believe in light particles. What about interference patterns, which can be proven in any school laboratory? What about the radio waves that the world is now using for long-distance communication? No, Bohr is certain: light and other electromagnetic radiation is made up of waves. And that means they can't be made up of particles. It may be that it can sometimes help to imagine them as little packets, but that's just a useful way of thinking about them.

However, Bohr's lecture at the Physical Society is not purely about the nature of reality, but also about prestige. Albert Einstein is in the audience. In consideration of him, Bohr avoids the question of the nature of radiation. Bohr was extremely impressed by Einstein's work in 1916 on the spontaneous and stimulated emission of radiation, and on the transmission of electrons between energy states. Einstein had made progress where he had been stuck, and had demonstrated that atoms behave according to the rules of chance and probability.

Einstein continues to worry about the fact that he is unable to predict when and in which direction light quanta will be emitted as an electron jumps from a higher to a lower energy state. However, he remains confident that he will find a way to repair this damage to the principle of causality. Now, Bohr contradicts Einstein in his lecture by arguing that there is no way to precisely predict the time or direction of such emissions, and that there never will be. The two men who admire each other so much now find themselves in opposing camps.

Over the next few days, as they stroll through the streets of Berlin or enjoy a meal at Einstein's home, each tries to persuade the other to switch sides.

Albert Einstein had not previously met Niels Bohr, who is six years his junior, but he has been an admirer of Bohr's since reading the Dane's first work, published in 1913, on the structure of the atom and spectral lines, in which Bohr brilliantly and ruthlessly applied 'quantum conditions' to Rutherford's planetary-system model of the atom. Einstein was later to say, 'That this insecure and contradictory foundation was sufficient to enable a man of Bohr's instinct and tact to discover the major laws of the spectral lines and of the electron shells of atoms together with their significance for chemistry appeared to me like a miracle—and appears to me as a miracle even today.'

When Bohr departs Berlin, he leaves a contented Einstein behind. 'It is not often in my life that I've met someone whose mere presence gives me such pleasure,' Einstein writes to Bohr after his departure. 'I'm now studying your great works and—whenever I get stuck—I have the pleasure of seeing your friendly, boyish face before me, smiling and explaining.' And Einstein wrote to his friend Paul Ehrenfest in the Dutch university city of Leiden, 'Bohr was here and I am just as keen on him as you are. He's like a sensitive child and goes about this world as if hypnotised.'

Bohr is equally enchanted by Einstein, but has a less flowery way of expressing it, replying to Einstein in his awkward German, 'It was one of the greatest experiences for me, which I have ever had, to meet you and to speak with you, and I cannot say how grateful I am for all the friendliness you showed towards me during my visit to Berlin. You don't know how great a stimulation it was for me to have the long-awaited opportunity to personally hear from you your views on questions I have concerned myself with. Never shall I forget our conversations on the way from Dahlem to your house.'

In August of the same year, Einstein pays Bohr a return visit, stopping off in Copenhagen on his way home from a trip to Norway. The two men will not spend such a peaceful time together again in the coming years.

Although Einstein learns no new facts from Bohr, he does learn something from him: 'Mainly how you approach scientific things intuitively.' Einstein recognises two kinds of physicists: 'principle-pedants' and 'virtuosos'. He considers himself and Bohr to be principle-pedants—that is, thinkers who delve deeply into a question in search of its basic principles. Max Born and Arnold Sommerfeld are virtuosos; they compose equations, but have no interest in philosophising about them. 'I can only further the mechanics of quanta,' writes Sommerfeld to Einstein. 'You must do the philosophising.'

One of the foremost reasons why Einstein admires Bohr so much for his theory of the atom is because Einstein himself had failed to come up with one. After his tremendous outburst of creativity in 1905, Einstein was left without an intellectual purpose. How could he surpass what he had already achieved? What might now attract his interest? Okay, there were those mysterious spectral lines, but Einstein can see no way to explain them with the knowledge physicists possessed at the time, so he leaves them unexplained. Instead, he returns to his theory of relativity—and formulates the equation $E=mc^2$. A few years later, he returns to the problem and believes he has almost found the answer. 'I currently have great hopes of solving the radiation problem,' he writes to a friend. 'And without using light quanta. I'm hugely curious to see how the thing turns out.' Almost as an afterthought, he adds, 'We should give up the energy principle in its current form.' Einstein considers the radiation problem to be so important that he's willing to abandon the principle of the conservation of energy. But even at that price, he doesn't find a solution. A couple of days later, a rather sheepish Einstein reports, 'Failed again to find a solution to the radiation problem. The devil is playing a cruel trick on me.'

Then Bohr succeeds where Einstein had failed. He comes up with an explanation for the spectral lines of hydrogen. For Einstein, it's 'like a miracle'. 'His must be a mind of the very highest quality, extremely critical and farsighted, which, however, never loses sight of the big picture,' Einstein enthuses about Bohr. In the summer of 1916, Bohr's model of the atom inspires in him 'a brilliant idea', as he calls it himself, for describing the emission and absorption of light in an atom. This idea leads him to a simple derivation of Planck's Law; not any old derivation, but '*the* derivation', according to Einstein. Now he is certain: light quanta really do exist.

But this realisation comes at a price. In order for Einstein to develop Planck's equation from Bohr's atomic model, he has to abandon classical physics' strict principle of causality. He considers three processes by which an electron can jump back and forth between lower and higher energy states in an atom. In the case of 'spontaneous emission', the electron jumps down to a lower state, and emits a light quantum as it does so. It can also do the opposite, jumping up a level and absorbing a light quantum in the process. In the case of 'stimulated emission', a light quantum nudges an excited electron, which consequently jumps down a level, giving off another light quantum. This is the process that will later form the basis for laser technology: Light Amplification by Stimulated Emission of Radiation.

The strange thing about these transitions is that they don't always have a cause. Sometimes, they occur completely spontaneously. Or they don't take place, even though the cause is there. All Einstein can do is calculate the probabilities of them occurring—in a similar way to Marie Curie with the phenomenon of radioactive decay. He's concerned about the principle of cause and effect that has governed physics since ancient times. Even Aristotle argued that there is a fundamental source of becoming in everything, that everything tends towards some end, or form: 'Everything that is generated is generated by something and from something and becomes something.' But electron jumps don't need a fundamental source of becoming; they can arise without cause.

Four years later, Einstein is still concerned about this. 'The causality thing still plagues me a lot,' he writes to Max Born in a letter dated January 1920. 'Is the quantum-like absorption and emission of light ever conceivable in the sense of the condition of complete causality, or is a statistical residue left? I must admit that I lack the courage of conviction here. But only very reluctantly do I give up *complete* causality.' Einstein is indecisive. He wants to move towards a new physics of quanta, but he is still unwilling to let go of classical physics.

Now, in 1920, Einstein is going through another creative crisis, which today would probably be described as a midlife crisis. He's already come up with the concept of light quanta, formulated his theories of both special and general relativity, been through a difficult divorce, and survived a war and several serious medical conditions. He still refers to himself as 'quite a sturdy fellow', but he now has to follow a strict diet. Now, at the age of forty-one, he's taking stock, asking himself whether the future has anything more in store for him, or whether that was it.

In the two years following their meetings in Berlin and Copenhagen, both Bohr and Einstein continue to wrestle with light quanta—each on his own. And it takes its toll on both. 'It is actually good that I have so many distractions,' Einstein writes to Paul Ehrenfest in March 1922, 'because otherwise the quantum problem would have driven me to the insane asylum long ago.' And in a letter to Arnold Sommerfeld in April of that year, Bohr laments, 'I have often felt myself scientifically very lonesome, under the impression that my effort to develop the principles of the quantum theory systematically to the best of my ability has been received with very little understanding.' It is time for the two greatest minds to argue again.

GÖTTINGEN, 1922

A son finds his father

Göttingen, on a sunny afternoon in June 1922. Two men are engrossed in conversation as they hike across the Hainberg Hills. Even from a distance, it's plain to see how different the two are. As one marches on with energetic strides, he has to keep stopping to let his companion catch up. The other moves as if he were considering every step carefully before taking it.

The elder of the two men is in his late thirties, his hair already beginning to turn grey. He's wearing a plain suit. He holds his head tilted to one side, and has a serious-looking face with a high forehead and prominent eyebrow ridges. He paces thoughtfully on, speaking German with a strong Danish accent. The other man could be his son, barely more than half his age at twenty, and looks even younger with his short blond hair, bright-blue eyes, and boyish face. He's clearly used to hiking.

The way the two men are walking together, they could indeed be father and son, or they could be old friends. But, in fact, this is the first time they've met.

Niels Bohr, the elder of the two, will be awarded the Nobel Prize for Physics in just a couple of months' time. He is visiting Göttingen in order to deliver a series of lectures to share his knowledge of the atom.

Visiting Germany is not an easy thing for Bohr in these post-war years. Denmark, which had remained neutral in the war, is now in dispute with Germany over parts of the territory of Schleswig on their mutual border. Travel is difficult. The reparation payments imposed on Germany after the war mean coal is in short supply, and the little that is available is of bad quality. Trains have to run slowly, and sometimes stop for hours on the open tracks when they run out of fuel.

Bohr did not have to take these difficulties upon himself. He no longer needs to travel in order to learn physics from others. On the contrary, they now come to him to learn. The Universitets Institut for Teoretisk Fysik, known as the Bohr Institute for short, had opened in Copenhagen on 3 March 1921. The growing Bohr family has moved into a seven-room apartment on the ground floor of the new institute building by the beautiful Fælledparken gardens. The Bohr Institute is an oasis of peace in a war-ravaged and crisis-rocked Europe.

This is a bleak but relatively calm time for Germany. The reparation payments are causing suffering among the population, which is exacerbated by the global economic crisis. At least there's no war on, and inflation is not so rampant that people have to cart money around in wheelbarrows just to pay for bread and milk. Most people have just about enough food to stave off starvation. Just days after Bohr's meeting with Heisenberg, the Jewish German foreign minister, industrialist, and writer, Walther Rathenau, is shot dead by extreme right-wing students—a portent of the Nazi terror to come.

The Munich physics student Werner Heisenberg also has enough food to avoid starvation, but not enough to stave off hunger every day. Although his family is among Munich's better-off, they cannot afford to pay for their highly gifted son to travel to Göttingen. Heisenberg's doctoral supervisor, Arnold Sommerfeld, pays for his train ticket out of his own pocket, and arranges for Heisenberg to stay with friends of his.

At a time like this, Bohr's trip to Germany is also a political statement. Like Einstein, he despises militarism and Germany's aspirations to become a major power. But he also resists efforts by some of his fellow physicists to exclude German scientists from the international community. Revenge is not a path to peace.

Soon after the end of the First World War, Bohr had re-established his contacts to Germany, which eventually have led to his trip to Göttingen for this 'Bohr festival'. That is the name given to Bohr's series of lectures, in reference to Göttingen's Handel Festival, which is taking place at the same time. More than a hundred physicists, both young and old, both theoretical and experimental scientists, have travelled from all over Germany and the rest of Europe to hear from Bohr's own mouth his view of the structure of the atom. They include such luminaries as Otto Hahn, Lise Meitner, Paul Ehrenfest, Hans Geiger, Gustav Hertz, Georg von Hevesy, and Otto Stern.

Bohr is able to explain the order of the elements in the periodic table by the arrangement of electrons around their atomic nucleus. He describes 'electron shells', which surround the nucleus like the layers of an onion. Each shell has space for a specific number of electrons. Elements with similar chemical properties have the same number of electrons in the outermost shell of their atoms, Bohr explains. Chemistry has now become physics.

In doing this, Bohr brings to light a numerical harmony in nature that has never been seen before. According to his model, the eleven electrons of a sodium atom are configured in three shells, with two electrons occupying the lowest shell, eight on the next, and one on the outermost shell. A caesium atom's fifty-five electrons are arranged on six shells containing two, eight, eighteen, eighteen, eight, and one electrons, in ascending order. Since the outermost shell of both contains a single electron, the two elements have similar chemical properties. All this, and much more, is predicted by Bohr's atomic model. It predicts that the as-yet-undiscovered element at position 72 should be similar to zirconium and titanium, which appear above it in the same column of the periodic table. It should not be similar to metals next to it, which make up the 'rare earths'.

(As it happens, shortly after Bohr predicts the properties of this undiscovered element in Göttingen, he is shocked by news from France. A group of scientists in Paris claim to have proven experimentally that element 72 is one of the rare earths after all. This news unsettles Bohr. At first, he doubts his own results, and then those of the French team. His friend the Hungarian chemist George de Hevesy and the Dutch physicist Dirk Coster conduct an experiment of their own in Copenhagen. They produce a larger amount of pure element 72 and are able to disprove the French results. The element is similar to zirconium rather than to the rare earths. On the request of the scientists who discovered element 72, it is soon given the name 'hafnium', from 'Hafnia', the Latin name of Bohr's native city, Copenhagen.)

Not everyone gathered in the auditorium in Göttingen is impressed by Bohr's feat. Where are his derivations, his equations? Where is the hard maths? But everyone is impressed by his ideas. Bohr is pleased. 'My entire time in Göttingen was a wonderful and educational experience for me,' he writes after returning to Copenhagen, 'and I cannot say how happy the friendliness I received from all quarters made me feel.' Bohr no longer feels alone, underappreciated, and misunderstood—the feelings that had taken root in him in the past few years are now somewhat assuaged. In his quest for a quantum theory, he is in the process of uncovering the innermost workings of the world. And hardly anyone seems to have noticed. But Bohr thrives on feedback. He's no self-sufficient genius like Einstein. He suffered under the isolation forced on Europe's physicists by the circumstances of First World War, even though he was able to spend the war years on the safe side of the Danish border.

On the morning of this summer's day, Bohr gives his third lecture. The seats at the front of the auditorium, flooded by the light of the summer sun, are reserved for the dignitaries of Göttingen's scientific community.

Heisenberg has to follow the talk from the very back of the hall, from where Bohr's gently spoken words are barely audible. In his typically artless way, Heisenberg dares to raise his hand, stand up, and express his doubts about Bohr's ideas. The lecture hall falls silent; heads turn. 'That's not correct,' Bohr hears the young German say. 'I did the maths.'

Spectral lines are a favourite topic of atomic physicists. When white light is passed through the vapours of various elements and then split with a glass prism, a pattern of black lines can be seen in the resulting spectrum, which are characteristic of the element in question. Physicists can use that pattern of spectral lines to clearly and reliably identify which element they are dealing with. But where do those lines come from? The answer must be connected to the structure of the atom—and that's the puzzle that physicists are trying to solve.

Bohr claims his atomic model means he can now explain the splitting of spectral lines in electric fields, a phenomenon known as the 'quadratic Stark effect', after the German physicist Johannes Stark, who discovered it a few years earlier. Or, more precisely, someone else can explain it. Bohr prefers to leave such painstaking tasks to his co-workers. In this case, it falls to his Dutch assistant, Hendrik Kramers. He has carried out the calculations for the interaction of atoms, as Bohr imagined them, with light in electric fields, and published the results in a research paper. Having had to study it in Sommerfeld's seminar, Heisenberg is familiar with that work, and has discovered errors in it.

Calculations are just not Bohr's thing. He acknowledges Heisenberg's discovery of a problematic matter, and reacts with magnanimity. After the lecture, he invites Heisenberg to take a walk with him.

As the two men head for the Hainberg Hills, Bohr cuts through the small talk and addresses the issue directly. He tells Heisenberg his now-nine-year-old atomic model should not be taken too seriously. What should be taken seriously, though, is the question of why atoms are so stable, and why the atoms of a given element are all precisely the same and remain so through all chemical, and many physical, processes. It seems like a miracle, Bohr philosophises, that traditional physics is unable to explain. So it must be time for physics to change.

Heisenberg can hardly believe his ears. Has Bohr just questioned his own atomic model, which physicists around the world have now adopted, and which is now taught at universities and displayed in museum exhibits? Indeed, he has. And more. Bohr doubts not only his own atomic model, but also the very idea that atoms can be understood in terms of visual models at all. Atoms as miniature solar systems may be a nice idea, says Bohr, but such images can only be a mental aid at best. At worst, they can instil in people the false feeling that they have understood something that science hasn't fathomed at all. The key question is how atoms manage to remain so stable through all the collisions and chemical reactions they are subject to, and why two atoms of the same element are always absolutely identical. It's a complete mystery from the point of view of classical physics. No mini solar system could remain stable through all the impacts and chemical reactions atoms undergo. Indeed, no system known to physics could do that.

Heisenberg, the high-flyer who is used to always being the first to understand everything in Sommerfeld's Munich seminar, listens intently, interjecting with occasional questions. He would love to know what quantum theory really means, he says. Beyond all the calculations, predictions of spectral line patterns, and quantum numbers, what does quantum theory say about the very nature of physical reality? What do all these strange equations mean?

'Meaning ...' replies Bohr. 'There are many ways language can mean things,' he says, as he waxes philosophical. 'If a single speck of dust is made up of billions upon billions of atoms, how can we reasonably speak of it as something small?' When it comes to atoms, language can be used only as in poetry. Just as a poet is not so concerned with facts, but rather with images and mental connections, the purpose of models in quantum physics must be to capture as much as possible of what we can know and say about atoms with our inadequate modes of thinking and means of expression. Why should we expect that the way of thinking we have developed for a world full of people, trees, and buildings will also be suitable for the world of atoms? 'The images we create for ourselves of atoms are developed, or, if you prefer, conjectured, on the basis of experience, not as a result of any theoretical calculations,' says Bohr. 'I hope those images describe the structure of the atom as well—and only as well—as is possible in the clear language of classical physics. We must be clear that language can only be used for this in a way similar to its use in poetry, where the aim is not to describe facts precisely, but to create images and establish mental connections.'

This insight is difficult for Heisenberg to take seriously. A couple of decades earlier, the Viennese physicist Ludwig Boltzmann vehemently argued that atoms are not just the abstract flights of fancy they had been considered to be since the time of the ancient atomists, and nor were they mere metaphors; rather, they were concrete objects, smaller than the chairs we sit on, but just as real. Boltzmann took his own life in 1906. His desperate mental state was at least partly due to the fact that his fellow physicists refused to believe his theory that atoms were real. However, in the years since his death, more and more physicists have come to accept the atomic theory of matter.

And now here is Bohr, claiming that the smallest building blocks of matter are nothing more than a manner of speaking, after all: a nice, but ultimately inadequate, figure of speech. Indeed, that is Bohr's basic argument, but with a twist that the earlier debate about atomism did not have. Bohr is not at all denying the existence of atoms, but he is saying that physicists can't ever hope to be able to describe their true nature. Perhaps atoms have no 'true nature'. Our conventional intuitions about the physical world, about matter, about things and their location and movements, simply break down at the smallest scale. But those are the only intuitions we have, and it is through them that we have learned to understand the world. We are not capable of simply throwing them overboard.

What Bohr is calmly explaining is disturbing for Heisenberg. Until now, all the formulae he debates with his teacher Sommerfeld and his fellow scientists have been mere tools to help make calculations and predictions that can then be tested experimentally. Now he is being told he should question the very nature of the world those predictions are supposed to describe. He recalls reading *Timaeus* on the roof in Schwabing three years earlier as Prussian government troops were engaged in battle with the Munich revolutionaries. The idea is now closer to him than ever: decoding the basic building blocks of the world by searching for Truth and Beauty.

'So what does quantum theory actually mean?' Heisenberg asks of Bohr. 'What kind of a world lies hidden behind all the clever calculations, spectral lines, and quantum numbers? What is the physics behind the formulae?' Bohr doesn't have a definitive answer for him, and certainly not one that could be described as simple. No, the classic models of the atom simply can't be correct. But they're not completely wrong, either. They are the best conceptual aids we have. The trick is to find models that account for as much as possible of what we can confidently say about atoms. One model alone will not be enough; multiple models will be needed, each complementing the others, but also contradicting them. An electron can be thought of as a particle or as a wave. Both are true, but not completely. In some aspects, electrons behave like particles; in others, they behave like waves. Our intuition may well baulk at this dual nature of the electron, but that's just the way the world is.

After fortifying themselves at the Rohns coffeehouse, the two physicists climb to the top of the Hainberg, which offers a view across the entire town. 'Do humans even have the capacity to comprehend that nature?' asks Heisenberg. 'Is there no prospect at all of our really understanding atoms?'

'Oh yes, there is,' replies Bohr, 'but in doing so we will come to understand for the first time what the word "understand" means.'

This way of talking about physics is entirely new to Heisenberg. Arnold Sommerfeld, the wiry little professor from Munich who had been Heisenberg's teacher, a figure from old Prussian times with his moustache and duelling scars, a teacher and mentor to countless physicists of the new generation, had always stressed that it was a physicist's job to carry out calculations and experiments, and to leave philosophising to others.

Heisenberg begins to realise that Bohr's way of thinking is completely different from that of almost all the other physicists gathered here in Göttingen, the stronghold of mathematical physics. Mathematical wizards and skilful experimenters abound. Bohr's strength lies elsewhere: in his intuition. He intuits the way the world is structured. This Dane doesn't do calculations; he muses, wrestles with words—just like a poet. Heisenberg is later to call him 'the only person who understands anything of physics in the philosophical sense'.

Where Heisenberg sees formulae, Bohr sees phenomena. Heisenberg senses that Bohr's insights do not come to him through logical deduction or by solving differential equations, but through 'intuition and inspiration', as Heisenberg would later put it. 'He chose his words very carefully, much more carefully than Sommerfeld usually did. And almost every one of his carefully formulated sentences revealed a long chain of underlying thoughts, of philosophical reflections, hinted at but never fully expressed. I found this approach highly exciting.' And now, as Bohr tells Heisenberg of his intention to decipher the atom, he speaks of it as if it were the task of a poet, like trying to find the words to say something that has never been said before.

For Heisenberg, it's as if he's learning about physics for the first time, all over again. In the three hours that walk takes, 'my real scientific development began', Heisenberg would later recall. And a friendship also begins between Niels Bohr and Werner Heisenberg that will eventually produce some of the most important steps in the formulation of a new quantum theory. It will endure for nineteen years before it fractures.

By the time the two men leave the Rohns coffeehouse and set off for the summit of the Hainberg, Bohr has already recognised his companion's extraordinary talent. He's impressed by Heisenberg's unquenchable thirst for knowledge. Bohr enquires about Heisenberg's plans, and invites him on a research visit to Copenhagen, even offering him the prospect of a scholarship. Heisenberg has certainly never expected such a commendation. Here's the chance to go to Copenhagen, to the great Bohr, his ultimate scientific hero! And the city is important to him in another, very particular way: Copenhagen is also the preferred destination of his great rival, Wolfgang Pauli.

Wolfgang Pauli and Werner Heisenberg are Arnold Sommerfeld's two most gifted students, but, being one-and-a-half years older, Pauli has always been one step ahead of Heisenberg. Pauli was already in Sommerfeld's seminar when Heisenberg began his studies. Sommerfeld charged Pauli with grading Heisenberg's assignments and advising him on which lectures and seminars to attend. While Pauli had received his doctorate *summa cum laude* the previous autumn, Heisenberg only passed his exam with great difficulty. Having avoided war service due to a weak heart, Pauli had just taken up his first academic position in Hamburg. He's considered the *wunderkind* of German physics.

The two were close enough in age to bring out the best in each other, but not close enough to become friends. Pauli liked to party the night away, drinking, getting into trouble, and sleeping till noon. Heisenberg preferred to be out in the countryside, heading for the mountains through the morning dew. Pauli was in the habit of greeting Heisenberg with, 'Good morning, you prophet of nature!' 'Good afternoon' was often Heisenberg's wry response. Pauli never missed an opportunity to call Heisenberg a *dummkopf*. 'That was a great help to me,' Heisenberg will later say.

Now he might have the chance to go to Copenhagen and study under the grand master of quantum theory personally, before Pauli does so. But old Sommerfeld has other plans for Heisenberg. He is first to continue his studies in Göttingen under Max Born. Pauli spends the winter semester doing the same — and is then allowed to join Bohr in Copenhagen. Once again, Heisenberg has to bide his time, but will eventually overtake Pauli. In 1932, at the age of thirty, Heisenberg will win the Nobel Prize for Physics, while Pauli won't receive his Nobel Prize until 1945, when he's forty-five years old.

One figure was missing from the Bohr Festival: Albert Einstein. He's currently fearing for his life. The political mood in Germany is becoming increasingly extreme. Nationalist newspapers are openly calling for the murder of Walther Rathenau, who has been foreign minister for just a few months. On 24 June 1922, two days after the end of the Bohr Festival, Rathenau is shot and killed in broad daylight by extreme right-wing assassins as he is being chauffeured from his residence in Grunewald to his ministry. His is the 354th political assassination by the far right since the end of the war. Left-wing extremist have killed twenty-two people in the same period. Einstein knew Rathenau well. The two men often discussed politics together. Einstein is one of those who warned Rathenau not to take on such an exposed ministerial position, and he's now afraid of becoming number '355+x' himself. Einstein is well aware that he is on the reactionary death squadrons' hit lists, and reacts by doing what he unsuccessfully tried to persuade Rathenau to do: he withdraws from the public eye, cancels his classes and public-speaking engagements, and ends his work with the League of Nations' International Committee on Intellectual

Cooperation. 'The situation here,' he writes in his explanatory letter to Geneva, 'is such that Jews are well advised to exercise restraint in all public matters.' He even considers giving up his position at the Prussian Academy of Sciences and taking a job in the patent office in Kiel. He is no longer simply a physicist, but has become both an icon of German science and a living symbol of Jewish identity—a potentially dangerous combination.

This difficult situation at least leaves Einstein with the time to read Bohr's work, and he is enraptured by it. 'This is the highest form of musicality in the sphere of thought.' And, indeed, Bohr's work does contain as much art as science. He gathers information from diverse disciplines—spectroscopy, chemistry—and, layer by layer, atom by atom, assembles the entire periodic table like a great jigsaw puzzle. This is all built upon Bohr's conviction that, although the microcosmic realm of atoms is governed by the rules of quantum mechanics, everything that is derived from those rules must be consistent with the phenomena we observe in our macrocosmic world, which is governed by the laws of classical physics. He calls this basic tenet the 'correspondence principle'. It forbids all concepts from the atomic world that do not correspond to the laws of classical physics in the macroscopic world. With this trick, Bohr manages to bridge the gap between the seemingly irreconcilable worlds of atomic and classical physics. Some critics consider the correspondence principle to be a 'magic wand that doesn't work outside of Copenhagen', as Bohr's assistant, Hendrik Kramers, would later say. Some wave this magic wand without success. But Albert Einstein recognises a fellow magician of his own kind.

However, the greatest recognition of the year still awaits Niels Bohr. In October, the Nobel committee awards him the Prize for Physics for his work on the structure of the atom.

A constant stream of congratulatory telegrams lands on his desk. Einstein also receives the Nobel Prize for Physics, belatedly for the year 1921, for his explanation of the photoelectric effect. But Einstein is not at home when the telegram from Stockholm arrives at his Berlin apartment. He and Elsa are setting out on a lecture tour, and on 7

October they board a ship in Marseille, bound for Japan. It's a welcome escape from the toxic atmosphere in Germany. Assassins don't tend to strike on ships. Einstein is delighted to discover that their fellow passengers are almost exclusively British and Japanese, calling them 'a quiet, fine company'. This could be a relaxing time for Einstein, who has packed many books and a little work in his suitcase, if it weren't for his delicate stomach and its adverse reaction to the swell of the sea. The Einsteins make stops in Colombo, Singapore, Hong Kong, and Shanghai. In some ports, they are greeted with a rendition of the German national anthem, which is uncomfortable for Einstein, as it reminds him of the unpleasantness he wanted to escape. On 17 November, the Einsteins disembark in Kobe and travel around Japan by train. Their journey takes in Hiroshima, the city that will be reduced to ashes twenty-three years later with the help of Einstein's equations. The Einsteins eventually return to Berlin in March 1923.

The Nobel Prize awards ceremony takes place in a snowy Stockholm on 10 December 1922, without the presence of Einstein. All doesn't go well for Bohr on this auspicious day. He leaves his lecture notes in his hotel room by mistake, and has to improvise. However, the mishap turns out to be a blessing, as it means his lecturing style is clearer and more entertaining than some audience members will have been expecting—until someone fetches his notes from the hotel. Then, to his own relief, but to the chagrin of the audience, he returns to his usual style of mumbled, read-aloud lecture notes.

The German envoy, Rudolf Nadolny, accepts Einstein's prize on his behalf, following a dispute with diplomats from Switzerland, which claims Einstein as its own. This prompts the Berlin Academy to telegraph to Stockholm: 'Einstein is an Imperial German.' At the post-awards banquet, Nadolny raises his glass and proposes a toast 'to the friends of my nation, which has once again produced from among its midst one who has achieved success for the entire human race'. Following the spat over Einstein's nationality, Nadolny expresses the hope 'that Switzerland, which offered a home and workplace to the scholar for many years, will also share this joy'.

Although Einstein had renounced his German citizenship in 1896 and assumed Swiss nationality five years later, he gained the status of a German civil servant when he took the position at the Prussian Academy of Sciences, which automatically made him a German citizen again—although he didn't even know it himself. Einstein doesn't care about such matters. Belonging to one nation or another is about as significant to him as deciding which insurance policy to sign, he is recorded as saying. But his stepdaughter Ilse asks the Nobel Committee to send Einstein's medal to the Swiss embassy in Berlin, since he is a Swiss national. The committee solves the problem by sending the Swedish ambassador in Germany to present Einstein with the medal personally.

MUNICH, 1923

A high-flyer almost crashes

Munich, 1923. Physics is far from almost everybody's mind in Germany as the country continues to be crippled by reparations payments following its defeat in the First World War. The economy is buckling under rampant hyperinflation. Fear of a communist takeover is also rampant. Many people are afraid that Russia's Bolshevik revolution will spread to Germany, and that farmers will be squeezed dry and factory owners' properties seized. France further fuels nationalism in Germany by insisting on the ruinous reparations payments.

The first German republic is struggling to survive. That's an opportunity for anyone intent on destroying it. In Munich, the painter Adolf Hitler tries to seize power in a coup. On 8 November 1923, Hitler storms a memorial event marking the end of the First World War. Together with Erich Ludendorff, a veteran of the war and far-right leader, he marches on the Feldherrnhalle beer hall. They intend to continue on to Berlin to topple 'the corrupt Weimar regime'. The coup attempt fails. Hermann Göring, the First World War fighter pilot ace, escapes, but Hitler is imprisoned, while Ludendorff is acquitted. From that moment, Hitler and Ludendorff are rivals. Hitler becomes the leader of the far right.

The Munich professor Arnold Sommerfeld has already fled the gloom and doom of Germany, having accepted invitations from a number of American universities to undertake a lecture tour in the winter of 1922–23. He ends up spending a year at the University of Wisconsin in Madison. His stay abroad is motivated not only by physics, but also by money. In times of hyperinflation, even a respected German professor can find it difficult to survive financially, and Sommerfeld knows that the dollars he earns here on any given day will still be worth enough to buy things the next. He's also motivated by the privilege of watching the American scientific world flourish at first hand—a rare honour at this time, when German scientists are still international outcasts. Sommerfeld is always careful to avoid any kind of political discussion.

In the first few weeks of 1923, Sommerfeld hears of a discovery made by Arthur Holly Compton, an experimental physicist. Although still only thirty years of age, Compton has already been head of the Department of Physics at Washington University in St Louis, Missouri, for two years. Compton has managed to precisely measure the deflection of X-rays by atomic electrons. He has stated that his findings are exactly what the quantum model of light predicts, and far removed from the predictions of the wave model. Compton bombards various elements, including carbon and others, with X-ray light. The majority of the rays pass right through the target samples. But Compton is interested in the secondary radiation caused by the scattering of the X-ray light by electrons. Does the wavelength of the light change? Yes, he discovers, and in doing so the light and the electrons behave like tiny little billiard balls colliding and flying apart.

On 21 January, an excited Arnold Sommerfeld writes to Niels Bohr to tell him of 'the most interesting scientific thing I've seen in America'. Contrary to all the evidence, Compton has demonstrated experimentally that X-rays behave as quanta. Sommerfeld is still somewhat hesitant to take Compton's results as concrete facts, as they have not yet been subjected to the normal scientific rigour of reproducibility and review. Compton's work will not be published until May 1923, in *Physical Review*, the most important American physics journal. However, that publication is virtually unknown in Europe. Anyway, Sommerfeld is already convinced that physicists are soon to learn 'a completely new lesson'. 'After this, the wave theory of X-rays will have to be abandoned for good,' he writes to Bohr. Sommerfeld states that, 'This is probably the most important discovery that could have been made in physics as it currently stands.' For Bohr, it is a blow.

For Einstein, Compton's results are a confirmation. Writing in the liberal left-wing newspaper *Berliner Tageblatt*, he says, 'The positive results of Compton's experiment prove that radiation behaves as if it consisted of discrete energy projectiles, not only with regard to the transference of energy, but also with regard to the impact of the radiation.' Of course, Einstein has been saying for years that light must be made up of particles. On the other hand, light must also be a wave—physicists have known this ever since Maxwell's time, and it's the principle that electrical engineers use to build radios and similar devices. But how can both be true? 'So there are now two theories of light,' Einstein writes, 'both indispensable and—as one must admit today, despite twenty years of great efforts by theoretical physicists—without any logical connection.' Both the wave and particle, or quantum, theory of light appear somehow to be valid. Light quanta cannot explain the wave-like behaviour of light, including the phenomena of interference and diffraction. But there is no way to explain the Compton effect or the photoelectric effect without recourse to a quantum concept of light. Light has two natures: it's both a particle and a wave. And physicists are just going to have to learn to live with that.

When Sommerfeld returns from Madison in the spring of 1923, Heisenberg also returns to Munich from Göttingen to finish his doctoral thesis, which deals with hydrodynamics—a theoretical analysis of turbulence in fluids. A sound-enough topic, but far removed from the object of Heisenberg's fascination: quantum theory.

Sound, but boring. And Heisenberg always has a problem paying attention to things that don't interest him. Reluctantly, he enrols in a course of laboratory practice for beginners. It's run by Wilhelm Wien, the Munich professor of experimental physics who gave his name to Wien's displacement law. He's a master scientist in the laboratory, and it was his precise measurements of electromagnetic radiation that paved the way for Max Planck's quantum hypothesis of 1900. But Heisenberg is not interested in practical work, and the fact that the university does not have the equipment he requires gives him a convenient excuse to neglect the experiments Wien has assigned to him, and to return to pure theory.

Heisenberg is called to defend his thesis in a *viva voce* examination in 1923. He's expected to display a broad basic knowledge of physics, both theoretical and experimental, and not just in his areas of interest. It's a struggle. The sections on mathematics, theoretical physics, and astronomy go well. But then come the questions on experimental physics. Wilhelm Wien has no intention of simply waving the high-flying student through this part of the exam. When Wien hears how little practical experimentation Heisenberg has actually done, he begins peppering him with questions to test his knowledge of experiment design. He asks Heisenberg to tell him the resolving power of a Fabry-Pérot interferometer, an instrument consisting of two parallel mirrors to measure the frequency of light waves. Not only is this obligatory knowledge for Heisenberg's examination, but it was also covered by Wien himself in one of his lectures. But Heisenberg was not paying attention, can't quote the relevant equations, and, under Wien's stern gaze, tries to derive them from first principles on the spot. But he can't do it, and when Wien asks him about the resolving power of a microscope, Heisenberg simply doesn't have the answer. Wien then asks him about the resolving power of a telescope, and, once again, Heisenberg has no idea. Wien asks him how an electric battery works. And, once again, Heisenberg disgraces himself.

Wien is about to fail Heisenberg when Sommerfeld intervenes. He argues that Heisenberg's performance in the final theoretical physics exam was so outstanding that he can't be denied a doctoral title. It's a tricky situation. Sommerfeld wants to make sure his star pupil passes the exam, without compromising his colleagues' integrity. The professors reach a compromise. Heisenberg receives his doctorate, but with the second-lowest possible grade.

Not willing to be tarred with this ignominy for long, Heisenberg turns his frustration into determination, and sets about learning everything he can about microscopes and telescopes. This knowledge will serve him well in the future.

Arnold Sommerfeld wants to keep the newly titled Dr Werner

Heisenberg in Munich, and Max Born wants him back in Göttingen. The two negotiate over Heisenberg's academic future. In a lengthy letter, Born convinces Sommerfeld that Heisenberg belongs in Göttingen. He praises Heisenberg's extraordinary talent, modesty, enthusiasm, zeal, pleasant disposition, and popularity among the physicists in Göttingen. Following his *viva voce* exam, Heisenberg will be able to achieve his post-doctoral habilitation under Born in Göttingen, and then work as an assistant professor there. Sommerfeld is won over, and accepts Born's proposal.

Heisenberg returns to Göttingen in September 1923 to pursue his post-doctoral degree under Born. The two scientists work together to try to calculate the orbits of the electrons in a helium atom. In the winter semester, Born gives a lecture with the rather cumbersome title 'The Application of Perturbation Theory to Atomic Physics', in which he describes, in a completely new way, the jumps of electrons between energy levels in atoms. These leaps are now part of the theory. Born introduces the concept of discreteness, in other words of quantum leaps, into the theory, laying the foundations for a new quantum theory.

Unlike Niels Bohr, Max Born is neither a philosopher nor a visionary, but he understands Bohr, and has the ability to express his words in mathematical terms. 'Mathematics is cleverer than we are,' says Born. Without keen mathematical minds like Born's, Bohr would be helpless. And without a profoundly creative thinker like Bohr, Born's acumen would fall on barren ground. Bohr's creative mind spews forth thoughts that he struggles to express in words and equations. Born is Bohr's diametrical opposite: hesitant, cautious, forgetful, scatterbrained, and prone to illness. Mathematics keeps him connected to the world.

While Born and Heisenberg wrestle with electron orbits in Göttingen, French troops occupy the Ruhr Valley. Germany is in arrears with the reparations payments it has to make to France under the Treaty of Versailles. Workers' strikes have been spreading in the industrial Ruhr Valley, as miners protest the fact that they are forced to dig coal for the French while they have nothing with which to heat their own homes and fill their stomachs. But the French are unbending. The

German government responds by printing money to pay the workers, which only serves to further devalue the mark. At the outbreak of the First World War, four German marks bought one American dollar; by July 1922, a dollar cost 493 marks; by January 1923, that had skyrocketed to 17,792 marks to the dollar; and in November 1923, one dollar was worth an unfathomable 4.2 quadrillion marks. (Written in numbers, this figure is a 10 followed by fifteen zeros.)

Many people have to queue up several times a week to collect their wages in cash—piles and piles of paper money representing millions and billions of marks—which they then hurry to spend at the baker's, butcher's, or dairy shop to stockpile as much food as they can. They know that just one day later, their money will only be worth half as much. The formally sedate German bourgeoisie now has to fight for its survival. 'In this way,' Max Born writes, 'the entire middle class lost most of its property, which made it easy prey for political agitators.'

At Germany's universities, there's little time for science. Students buy food when they have money, and go hungry when they don't. Lectures are cancelled; lecturers and assistants face pay cuts.

Max Born escapes into his research, as far as his asthma and bronchitis allow. He increasingly withdraws from friends and colleagues, playing the piano alone and often working late into the night. 'I myself have absolutely no yearning for high physicist concentrations,' he writes to Einstein in August 1923, 'but would rather prefer to live and work quietly on my own. As of tomorrow, I shall play dead and won't be available. I actually don't have anything special planned. I am thinking, hopelessly as ever, about quantum theory and am searching for a recipe to calculate helium and the other atoms; but I'm not succeeding in that, either.'

In August 1923, Gustav Stresemann is appointed German chancellor. He changes the government's monetary policy, puts an end to the rampant printing of banknotes, and introduces a new currency. One new Rentenmark is worth several million old marks. 'One cannot remove a problem from the world by giving the independent variable a new name,' criticises the Göttingen mathematician David Hilbert, whom some consider infallible. But this time, Hilbert is wrong.

The Rentenmark not only has a new name, but it is also covered by corporate bonds. Prices recover, and the economy revives. And so does scientific activity. What remains is a deep mistrust of the young state among many people in Germany. In the spring of 1923, Max Born writes to Einstein, 'The madness of the French makes me sad because it strengthens nationalism over here and weakens the republic. I think a lot about how I could arrange for my son to be spared the fate of having to participate in a war of reprisal. But I am too old for America and, anyway, an even greater war madness than here has reigned over there.'

In Heidelberg, physics professor Philipp Lenard concludes his lectures for the 1923–24 winter semester with a panegyric on the imprisoned coup leader Adolf Hitler, and in 1924 he publishes a declaration entitled 'The Hitler Spirit and Science', in which he expresses his enthusiasm for National Socialism, and describes Hitler and Ludendorff as the embodiments of the 'cultural hero spirit' of 'Aryan' scientists such as Newton, Galileo, and Kepler. The publication goes on to claim that 'from time immemorial, from the crucifixion of Christ to the burning of Giordano Bruno and the incarceration of Hitler and Ludendorff, the same Asian people have always been in the background, working against these cultural heroes.' (In fact, Ludendorff had been arrested, but not convicted.)

COPENHAGEN, 1923

Bohr and Einstein take the tram

In 1922, Albert Einstein receives the 1921 Nobel Prize, but doesn't attend the award ceremony. And so it's not until 1923 that he delivers his Nobel lecture in Stockholm, in front of an audience of 2,000 people. Einstein is certain that most of them are just there to see him in person, not to listen to what he has to say. And so he's relieved to board the train that will take him from the Swedish capital to the man who he knows will certainly listen very carefully to everything he says, and contradict almost every word.

Bohr is waiting on the platform at Copenhagen station to greet Einstein. It's the first time the two men have seen each other for three years. Three years in which a lot has happened. The two professors immediately become engrossed in a conversation about physics. Oblivious to everything around them, they absentmindedly board the tram to Bohr's institute. Bohr has heard from Arnold Sommerfeld about Arthur Compton's groundbreaking experiment in the US. Einstein has also heard about the young American's work. Einstein is no longer alone in defending the concept of light quanta, but Bohr still refuses to believe in them. 'But, Einstein, you must understand …' insists Bohr in his Danish-accented German. 'No, no …' counters Einstein, and argues against Bohr's quantum leaps. 'But, but …' answers Bohr. Unmindful of the bewildered looks they're attracting from the other passengers, they miss their stop and travel far too far on the tram. 'Where are we?' asks Einstein. Bohr doesn't know. They get off and catch another tram going back the other way. And miss their stop again. 'We went back and forth many times on that tram,' Bohr later recalls. 'I can only imagine what people thought.'

Niels Bohr at his desk at the Institute for Theoretical Physics at Copenhagen University in 1930

COPENHAGEN, 1924

One last try

Niels Bohr is outraged. Arthur Compton's discovery has put him on the back foot, but he remains convinced that light is a wave. However, Compton is not the only one challenging Bohr's convictions. His own assistant, Hendrik Kramers, had the same idea about quantum collisions a few months earlier than Compton. He was very proud and excited to present his idea to Bohr.

Kramers, of all people! Kramers, who's been Bohr's most loyal assistant for years. He turned up at Bohr's door in 1916, his freshly passed physics degree in hand, keen to learn about the new quantum theory. He proved himself to be a talented student; a little unsure of himself, perhaps, but with a winning sense of humour. 'Bohr is Allah, and Kramers is his prophet,' Wolfgang Pauli once observed. And now the prophet is betraying his god by providing the most important clue to the existence of light quanta? 'No way!' is Bohr's reaction. He berates Kramers, swearing that there is no place in physics for a particle theory of light, remonstrating with him that he should not play fast and loose with the highly successful theory of electromagnetism, and telling him to toe the line. Bohr can be very convincing, even when he's being unreasonable.

Kramers toes the line. Under pressure from Bohr, he buries his discovery, destroys his notes, becomes ill with worry, and even has to spend a few days in hospital. Bohr has really twisted his mind. When Compton publishes the discovery that Kramers had already made, he joins his boss in resisting it.

Bohr knows they are in the minority, but he won't give up. Not yet. Together with Kramers and the twenty-two-year-old American theoretical physicist John Slater, who is currently in Copenhagen as part of a European tour, he makes one last attempt to save the wave theory. Like most physicists of his young generation, Slater has no aversion to light quanta. He's young and undogmatic. Why shouldn't light quanta and light waves be able to coexist? But he's now in Copenhagen, the stronghold of the anti-quanta resistance. Bohr and Kramers win him over to their cause. Within just three weeks, they write a research paper together—that's faster than any other of Bohr's publications. Bohr thinks aloud. Kramers takes notes, the best he can. Slater stands by, listening and marvelling, occasionally interjecting with a question.

By January 1924, their paper has already appeared in the *Philosophical Magazine*. It's filled with sentences that are typical of Bohr's enigmatic style:

> We will assume that a given atom in a certain stationary state will communicate continually with other atoms through a time-spatial mechanism which is virtually equivalent with the field of radiation which on the classical theory would originate from the virtual harmonic oscillators corresponding with the various possible transitions to the other stationary states.

All these notions—'virtually equivalent', 'time-spatial mechanism', and 'communicate'—have no clear meaning in physics. Einstein says of Bohr that he 'expresses his opinion like someone who is constantly searching and not like someone who believes he's in possession of the ultimate truth'. Einstein means this as a compliment, but it also exposes one of Bohr's weaknesses. He can rarely be tied down to one statement, always seeming a little evasive.

Bohr is prepared to abandon the basic law of the conservation of energy and the conservation of momentum in order to accurately describe the absorption and emission of light quanta in atoms. He wants to reduce these laws to the level of mere statistics: sometimes energy and momentum are conserved; sometimes they're not. As a defender of fixed laws of nature, Albert Einstein cannot agree with this. A law is a law, not a rule of thumb. But Einstein doesn't have a better suggestion.

Bohr's core idea is to sacrifice the laws of the conservation of energy and momentum in order to preserve wave theory. This law is a cornerstone of physics, and is an essential ingredient in Arthur Compton's argument for the effect that bears his name and that has now brought the quantum theory of light very much to the forefront. If the conservation of energy does not apply as strictly at an atomic scale as it does in the day-to-day world of classical physics, the Compton effect can no longer be taken as clear, unambiguous proof of Einstein's light quanta.

The three physicists in Copenhagen argue that the law of the conservation of energy has never been tested experimentally at the atomic scale. They consider that the question of whether and to what extent the law applies to the spontaneous emission of light quanta is one that still remains to be answered. Einstein assumes that energy and momentum are conserved in every single collision between a light particle and an electron. 'Why should it?' asks Bohr, and counters Einstein's assumption by arguing that energy and momentum are not necessarily conserved in each interaction, but only overall, statistically.

Bohr, Kramers, and Slater do not have a fully constructed model to counter Compton's precisely calculated scattering theory. They sketch out a possible theory in purely qualitative terms, and their research paper includes only one, extremely simple equation. They describe a new kind of 'virtual' radiation field that surrounds atoms, influences their absorption and emission of light, and transfers energy between them. The field acts as an energy buffer, so that, although the energy is not conserved in every single interaction between light and an electron, it is conserved overall over the long term. But where does this mysterious radiation field come from? Bohr, Kramers, and Slater have simply made it up.

At first glance, the BKS theory, as it becomes known after the initials of its three originators, appears to be a surprising piece of sleight of hand in the debate about the nature of light. But, in fact, it is an act of desperation that shows how vehemently Bohr rejects Einstein's quantum theory of light. Anxious to know what Einstein thinks of the suggestion, but lacking the courage to ask him directly, Bohr asks Pauli to do it. In September 1924, Pauli finally manages to meet with Einstein and get his opinion of the BKS theory. Einstein dismisses it as 'distasteful' and 'abhorrent'. Pauli bluntly relays to Bohr Einstein's dismantling of the theory with phrases such as 'completely artificial', 'something shady about it', 'offensive', and 'no logical connection'. Pauli agrees with Einstein, decrying what he calls 'the virtualisation of physics'.

Can it still even be called physics, if electrons that are irradiated with light can jump around inside their atoms like balls in a roulette wheel? Not according to Einstein's idea of physics. As he writes to Max Born: 'The thought that an electron subjected to a ray chooses of its *free volition* the instant and direction in which it wants to bounce away is intolerable to me. If so, then I would rather be a cobbler or even a casino employee than a physicist.'

Niels Bohr senses that he is at a fork in the road. This is the point at which his tenacity could easily degenerate into obstinacy. The point at which he could become the tragic hero who loses his way defending an untenable theory from the previous century. But it won't come to that. Just a year later, experiments carried out by Compton — now based at the University of Chicago — and Hans Geiger and Walther Bode at the Imperial Institute for Physics and Technology in Berlin show that energy and momentum are conserved when light and electrons collide. Einstein is right. Bohr is wrong, but he's not the kind of person who will sacrifice the truth to cling to his position in an argument. He suggests giving 'our revolutionary efforts as honourable a funeral as possible'. Bohr suspects that the truth about quanta is far deeper than anyone, even Einstein, himself, or anyone else, has ever imagined. He writes, 'One must be prepared for the fact that the required generalisation of classical electrodynamic theory demands a profound revolution in the concepts on which the description of nature has until now been founded.'

The glory of the Bohr Festival in Göttingen has now faded. After providing for Bohr's astonishing predictions about the spectral lines of hydrogen and ionised helium, his model reaches its limits when it comes to helium atoms with two electrons. What are the two electrons' orbits? Bohr can't answer that question, and Kramers can't help him. 'All attempts to solve the problem of neutral helium atoms have proven unsuccessful,' states Arnold Sommerfeld.

This is a low point for quantum physics. None of the scientists can see a light at the end of the tunnel, and they're all just stumbling around in the dark. 'It's a shame,' laments Werner Heisenberg. 'Once again, physics is very confused at the moment,' writes Wolfgang Pauli, 'at least, it's far too difficult for me, and I wish I was a movie comedian or something and had never heard of physics at all!' And Max Born states, 'Presently we have only a few vague hints as to the nature of the deviations that need to be imposed onto the classical laws in order to explain the atomic properties.'

Born spends the spring of 1924 confined to his bed with bronchial asthma due to pollen allergies. His colleague Heisenberg, also sniffling with hay fever, travels to Copenhagen for his first visit with Bohr.

Heisenberg arrives at a very busy time for Bohr. He's the father of five children and the head of a rapidly growing research institute. He would like very much to get to know Heisenberg better, but there is little opportunity for quiet conversation. So he suggests to Heisenberg that they take a hiking tour together through the countryside on the island of Zealand. The two men pack their rucksacks and hike north to Elsinore, the location of Hamlet's castle, then head back down the Baltic Sea coast to Copenhagen. In just a couple of days, they cover more than 150 kilometres, all the while talking about life and politics, and discussing the power of images and concepts in atomic physics. At one point on their journey, seeing a roadside telegraph pole in the far distance, Heisenberg picks up a stone and throws it at the pole, and hits it—against all the odds. Bohr thinks for a while, then says, 'If you had thought first about your aim, or about the correct angle of your arm and wrist, you wouldn't have had the least chance of scoring a hit. But since you were unreasonable enough to imagine that you could hit the target without special effort, you did it.' Sometimes, imagination is more powerful than reason. Speaking to one of his students after they returned from their hike, about the current state of quantum physics, Bohr says, 'It's now entirely up to Heisenberg to find a way out of these difficulties.'

Heisenberg spends two weeks with Bohr. Returning home, he excitedly brings with him the news of the BKS theory. Born is alarmed, but for other reasons. He doesn't like Heisenberg's obvious affinity with Bohr. He has barely managed to prise this young star of quantum physics, this 'kind, precious and very clever person who has grown dear to my heart', away from his beloved Munich, and now he has to fear losing him to Copenhagen. Bohr sends a letter to Born, asking him to let Heisenberg come and work with him, and Born is unable to refuse. He knows that almost anywhere in the world would currently be a better place to live in and carry out research than Germany, and he's planning to leave for a lecture tour in America himself. At this time, he can have no idea that he will be the one to discover the fundamental rule of electron roulette, upon his return from the United States.

PARIS, 1924

A prince makes atoms sing

One spring day in 1924, a parcel bearing a Paris postmark lands on Einstein's desk. A French friend is asking Einstein for his opinion of the enclosed work. Opening the typescript, Einstein reads the rather nondescript title, '*Recherches sur la théorie des quanta*' ('Research on Quantum Theory'). The work is not very long, is quite an easy read, and mentions Einstein's theory of relativity. The author of the paper, who wants to present it as his doctoral thesis and is unknown to Einstein, is already thirty-one years old. An eternal student? At least he has a fine name: de Broglie—an actual prince! Einstein starts reading, and is impressed. This prince has come up with a new way of looking at the nature of matter.

Prince Louis Victor Pierre Raymond de Broglie has not simply been dragging his heels at university—aristocratic family duties and the war both slowed down his academic career. He's from a grand old French aristocratic family, and has grown up with the expectation that he will follow in the footsteps of his forefathers. The House of de Broglie originally came from Piedmont in northern Italy, where its members distinguished themselves as military commanders and monastery patrons from as early as the twelfth century. For hundreds of years, members of the family have been statesmen, politicians, and officers in France. Scions of the house have been eminent military leaders, historians, and statesmen, including three marshals of France and two prime ministers. The French king awarded the line the title of duke, and the German Holy Roman emperor gave them the title of prince. Thus, Louis is both a French duke and a German prince.

Louis' story starts on 15 August 1892 in Dieppe, when he is born as the youngest of four surviving children. He and his siblings grow up with the luxuries and privileges of a French aristocratic family, although their parents are distant and cold. As befits children of the nobility, they are educated at home by private tutors. Relations within the family are very formal, even among the siblings. As a little boy, Louis regularly reads the newspapers and gives mock speeches to his family. At the young age of ten, he's able to recite the names of all the ministers in the government of the Third Republic.

There seems to be little doubt about it: Louis is destined to become a statesman, perhaps a parliamentary deputy like his father. Or even prime minister, like his grandfather. But when Louis is fourteen years old, his father dies. Louis grows up under the wing of his elder sister, Pauline, Comtesse de Pange, who loves and adores him. She describes him as a delicate boy, 'curly haired like a poodle, with a cheerful little face and mischief in his eyes'. His cheerfulness, his 'boyish exuberance', fills the cold, bleak rooms of the family chateau. 'He chattered incessantly at the dinner table,' she gushes, 'and even when admonished in the strictest terms to be silent, he was unable to check his tongue; his remarks were simply irresistible! In the solitude in which he grew up, he read a great deal and lived in a world of fantasy. He had an astonishing memory and could recite entire scenes of the classical theatre from heart with inexhaustible gusto. But he would tremble with fear even in the most harmless of situations: pigeons would scare him, he had a terror of both dogs and cats, and an attack of panic would sometimes befall him on hearing the sound of our father's shoes on the stair.'

Louis' elder brother, Maurice, becomes a surrogate father figure to him, although Maurice doesn't take after the rest of the family. Rather than joining the army, he goes into the navy, where he helps develop a wireless communication system for ships, and writes a paper on 'radioelectric waves'. Fired up by the hype around X-rays in the 1890s, Maurice decides to become a scientist, against the will of his father. He raves about radiation and electrons to his younger brother.

Maurice sets up a laboratory in one of the family's residences, in rue Chateaubriand, and it continues to grow until it takes up most of the building, including the guest bathroom and the stables. They now hum with the sound of transformers, connected by thick cables to vacuum tubes so as to generate X-rays.

In October 1911, Maurice travels to Brussels, where the first Solvay Conference is taking place at the Hotel Metropole. It's the most important meeting of scientists at the time. Participants include his French compatriots Marie Curie and Henri Poincaré, along with Max Planck, Albert Einstein, Ernest Rutherford, Arnold Sommerfeld, and Hendrik Lorentz, who are all there to discuss 'The Theory of Radiation and the Quanta'. Maurice has the honour of being named secretary of the conference. He produces transcripts of the discussions, prepares them for publication, and also passes them on to his brother, Louis: 400 pages full of the ideas and arguments of the world's leading physicists. Reading these transcripts is a turning point in Louis' life, and he then follows in his brother's footsteps, becoming a scientist and even resolving to '*épouser la science et elle seule*'—'to be married to science and science alone'. He breaks off his engagement to the daughter of a respectable family, bans bridge cards and the chessboard from his chambers, and purges his personal library of all history books. He cuts social contacts to a minimum. From now on, all that counts is science.

Pauline hardly recognised her darling little brother. 'The *petit prince* who was such a delight to me in childhood was gone forever. He now spent his entire time withdrawn in his little room, buried in mathematics books and bound by a rigid, never-changing daily routine. With shocking speed, he became an ascetic who lived such a monkish life that his left eyelid, which had always drooped somewhat, now almost completely covered his eye and so disfigured him that I felt sorry, since it accentuated his absent, effeminate nature all the more.'

In 1913, shortly before the outbreak of the First World War, Louis de Broglie recklessly volunteers for the Army Engineers Corps to complete his military service. During the war, he serves in the Signal Corps, working as a radio engineer relaying radiotelegraphs from his station beneath the Eiffel Tower. His duties include maintaining the radio equipment used to intercept enemy communications. Far removed from the trenches, it's a relatively comfortable post for a young man in the war, but still a tough time for the less-than-heroic Louis. He would later complain that his head never functioned as well after the war as it did before it.

After the war, Louis de Broglie maintains only one relationship. It's with a young artist by the name of Jean-Baptiste Vasek, a painter and collector of what he calls 'brutal art'. He gathers poems, sculptures, drawings, and paintings by psychiatric patients. He believes their visions to be the source of the myths of the future. Louis adores talking with his friend, but also the silences they share together, and falls in love with the artist. But, suddenly, Jean-Baptiste takes his own life. Leaving no explanation, the note he writes to his 'most beloved Louis' simply asks him to take care of and keep adding to his art collection. Setting aside his study of physics, Louis pours all his energy and his share of the family fortune into fulfilling his friend's dying wish. He tours the insane asylums of France and the rest of Europe, and organises a major exhibition with the title *La folie des hommes* (*The Madness of Mankind*). It's a success with the public, and a scandal among the aristocracy. When Louis reads a newspaper article mocking his friend's paintings, he's deeply hurt and shuts himself away in his rooms for months. His sister brings him food, but he leaves it outside the door, untouched. Pauline begins to suspect that her brother is trying to starve himself to death. She alerts her other brother, Maurice, who breaks the lock to Louis' door and storms into his chambers with five servants, rushing past all the works of art to the bedchamber, where he expects to find his brother in a desolate state. But instead he finds Louis well kempt and smartly dressed, his gaze clear, and a cigarette between his lips. He hands Maurice a stack of papers covered in equations, with the words 'I need you to tell me if I've lost my mind.'

During the war, Louis learned the value of classical electromagnetic wave theory, and his brother has told him about the controversial concept of light quanta. Like many scientists, he considers these two ways of interpreting light to be mutually incompatible. But he finds a way to approach this conundrum that has never been considered before.

In late 1923, 'after long reflection in solitude and meditation', de Broglie comes upon a simple but bold idea. Turning Einstein's argument for the photoelectric effect on its head, he asks the question: if light can behave like a stream of particles, can particles behave like waves in some respects?

It's a new, daring, and not very well-founded conclusion, and a leap in the dark. Until now, it has simply been taken for granted that particles are compact concentrations of mass, with absolutely no connection to waves at all.

Louis de Broglie simply combines Einstein's equation $E=mc^2$ and Planck's equation $E=hf$, and calculates a wavelength for each moving particle. The faster the particle is moving, the shorter its wavelength.

Is it more than just a mathematical trick? Is there an actual wave with the wavelength calculated by de Broglie? He doesn't know himself. He has simply tried it—and it worked, much to his own surprise. For an electron on the innermost orbit of an atom, according to Bohr's model—which even Bohr himself no longer believes in—de Broglie calculates a wavelength that is exactly the same size as the circumference of the orbit. For an electron on the second-lowest orbit, the circumference is equal to twice the wavelength. On the third-lowest orbit, the circumference is equal to three times the wavelength. And so on. That has to be more than just a coincidence.

De Broglie has made atoms sing. Like the fundamental frequency and the overtones of a violin or guitar string, the permissible orbits in Bohr's atom are precisely those that can accommodate a whole number of wavelengths of the electron's 'wave'. Are quanta nothing more than a kind of musical harmony?

De Broglie publishes his idea in two papers at the end of 1923. And no one takes any notice. Having thought the concept through even more thoroughly, he's confident and self-assured when it comes to defending his doctoral thesis. However, he's unable to persuade his examiners. The idea of electron waves seems too simple from a mathematical point of view, and too arbitrary from the point of view of physics. They can't prove de Broglie's calculations are wrong, but they fail to recognise their value to physics. Even de Broglie's doctoral supervisor, Marie Curie's former lover Paul Langevin, is unsure what he should make of the thesis. 'Looks far-fetched to me,' he confides to a colleague. But he recognises its potential importance enough to send a copy to his friend Albert Einstein to ask his opinion of it.

Einstein doesn't reply for several months. Langevin becomes increasingly anxious. Does Einstein think the thesis is so unworthy that it doesn't even warrant a response? Langevin asks again. This time, Einstein replies to his letter immediately. As a fan of simple, bold ideas, Einstein's response is short and to the point: 'Very impressive.' Einstein says de Broglie is on to something big, and has 'lifted a corner of the great veil. I think it is the first ray of weak light to illuminate these rules, the worst of our physical puzzles.' However, so far, very few appreciate what de Broglie has achieved.

Langevin takes heed of Einstein's comments. In November 1924, the Faculty of Physics at the University of Paris convenes to hear Louis de Broglie defend his doctoral thesis. Raising his nasal, aristocratic voice, and with a soporific tone, de Broglie begins to spout outrageous claims:

> Physics in its contemporary state contains false doctrines that exercise a dark influence on our imagination. For more than a century, we have divided earthly phenomena into two fields: atoms and particles of solid matter on the one hand, and the intangible waves of light, propagated through the sea of the luminiferous ether, on the other. But these two systems cannot remain separate; we must bring them together in a single theory, as only that can explain their many interactions. Our colleague Einstein has already taken the first step: twenty years ago he postulated that light is not only a wave, but also contains energy particles, and those photons, which are nothing more than amounts of energy, travel along with light waves. Many have doubted the veracity of this concept, while others have preferred to close their eyes so as not to see the new path he shows us. Let us not deceive ourselves, this is truly revolutionary.

> We are speaking here of the most cherished object of physics—light—and light not only allows us to see the forms of this world, it also shows us the stars that adorn the spiral arms of the galaxy and the hidden heart of things. However, this object is not single, but dual. Light can exist in two different ways. As such, it defies the categories into which we have attempted to divide the countless forms in which nature manifests itself. Light occupies two systems, as a wave and as a particle, it has two identities and they are as diametrically opposed as the two faces of double-headed Janus. Like that Roman god, it expresses the contradictory properties of the continuous and the dispersed, of the separate and the same.
>
> Those who reject this argue that accepting such a doctrine would mean renouncing reason. But to them I say, all matter possesses this duality! It is not only light that experiences this division, but every single atom that makes up the universe. The dissertation you now hold in your hands demonstrates that for every particle of matter—whether electron or proton—there is a corresponding wave that transports it through space. I know there are many who will contest my arguments and I do not hesitate to admit that they are solely a product of my own solitary reflections. I accept that they are bizarre in character and I also accept the punishment that may be meted out to me should they be wrong. But here today, I say to you, out of the deepest conviction, that all things can exist in two different ways and nothing is as fixed as it seems. The stone held in the hand of a child and aimed at the dim-witted sparrow on the bough could simply trickle through his fingers like water.

De Broglie ends his speech. The dumbfounded professors say nothing; they find no words to argue with his thesis. Louis de Broglie leaves the examination hall having received his doctorate. And five years later, he will receive a Nobel Prize for his work.

Einstein is so taken by the idea of waves of matter that he suggests, at their next congress in Innsbruck in September 1924, that his colleagues in experimental physics start looking for indications of the wave phenomenon in molecular radiation. Einstein sees de Broglie's waves of matter as a step towards re-establishing the classical order in physics that Bohr is currently undermining. That comes at a price, however: now there are two coexisting aspects of matter—waves and particles. How are they related? Are they related at all? Einstein does not have the answer.

One chance event in April 1925 shows that de Broglie is on the right track with his bold idea. Clinton Davisson, a thirty-four-year-old physicist working at the Western Electric Company in New York, is studying what happens when electrons are fired at different metals. One day, a flask of liquefied air explodes, smashing a vacuum tube containing a sample of nickel that was being bombarded by a beam of electrons. After the nickel oxidises on exposure to the air, Davisson cleans it by heating it up, without realising that the heat causes the tiny crystals in the nickel to reform into much larger ones, whose crystalline structure deflects the electron beam. Davisson is unable to explain why his results are now suddenly so different. He records them, publishes them, and then is astonished when colleagues explain to him what his measurements mean: electrons can behave like waves. The French prince's guess was right. Twelve years later, Davisson receives a Nobel Prize for his discovery.

HELIGOLAND, 1925

The vastness of the sea and the tininess of atoms

May 1924. Nature is in full bloom. And Werner Heisenberg, now twenty-four years of age, is suffering. His eyes sting, and his face is swollen and red. His nose is blocked, and, Heisenberg feels, so is his brain. Now, of all times!

Like many physicists at this time, Heisenberg is puzzling over the mechanics of electron orbits in atoms, how electrons can jump back and forth between orbits, and how those leaps cause the lines that physicists and chemists see with their spectroscopes. And he now has an idea that could explain the jumps; a bold idea that he senses could lead to a new theory of quanta. Sometimes the essence of genius lies in questioning the obvious. Unlike his colleagues, Heisenberg is prepared to abandon the laws that have governed scientific thinking since the time of Isaac Newton. He suspects that those laws lose their validity inside atoms. If he wants to find out what holds the smallest particles together at the deepest level, he will have to take an entirely new approach.

But genius also requires patient, hard work. The mathematical formalisation of his idea causes no great difficulties for Heisenberg. He needs to describe the position and velocity of an electron as a function of the wavelength of the atom it belongs to. But when he puts those functions into the equations of classical mechanics, the results are just an inextricable mess. Single figures quickly grow into whole series of numbers; the traditional rules of algebra generate pages and pages of equations. Heisenberg starts trying out different things. He plays around with Fourier series, gets bogged down, and begins to lose his patience. And on top of that, he's suffering from this massive hay fever attack that seems to be crippling his brain.

Heisenberg presents such a pitiful sight that his boss, Max Born, grants him special leave of absence. On 7 June, fleeing the pollen, he boards the night train to Cuxhaven, where he takes the ferry to Heligoland in order to recover in the sea air. Heligoland, named after the mediaeval German for 'holy land', is Germany's only island in the open sea. It's smaller in area than Berlin's Tiergarten park, and is eighty kilometres from the coast. The climate is so harsh that barely a flower blooms on the island, and hardly a tree can grow to any great height. During the First World War, Heligoland was a military outpost, but since the war it has become a holiday destination, beloved of tourists seeking peace and fresh sea air. Heisenberg finds lodgings, and on seeing him, his landlady suspects he's been beaten up — not an uncommon occurrence in Germany in the years since the war.

Heisenberg takes little with him: just a change of clothes, a pair of hiking boots, a copy of Goethe's *West-Eastern Divan*, and his calculations concerning electron orbits. His room is on the second floor, high above the island's red sandstone cliffs. During his week-and-a-half stay, he sits on his balcony, breathing deeply and looking out over the sea, walks on the beach, swims out to the neighbouring island, reads Goethe's verses, speaks to no one, and thinks — day and night. This is Heisenberg's happy place. He has always found refuge in nature — in the mountains, in the forest, by the water. Gradually, his mucous membranes return to normal, and his mind becomes unclouded. Back in Göttingen, Heisenberg has been too deeply mired in traditional thinking about physics, but here he has a wide, unhindered view all the way to the horizon. He thinks back on the words Niels Bohr once used to persuade him, as an avid mountain climber, of the magic of Denmark's flat, coastal landscape: 'Part of infinity seems to lie within the grasp of those who look across the sea.' In the solitude of Heligoland, Heisenberg now understands what Bohr was trying to tell him that day: how overly simplistic it is to

Werner Heisenberg in Chicago in 1929

imagine atoms as tiny solar systems with electrons revolving around the nucleus as the planets circle our Sun. Heisenberg extinguishes the Sun and smears the sharply defined orbits of the electrons, scattering them into amorphous clouds.

Thus he discovers a new way to approach the puzzle of quantum leaps. Someone in the Göttingen group—perhaps it was Wolfgang Pauli—once claimed there's only one way to find out what's happening inside an atom or any other similarly small system, and that's through measurement. You can measure the quantum state of an atom, then measure it again later to ascertain its new state. But what happens in between? Does it even make sense to speak of an 'in between'? The idea begins to form in the Göttingen group that the only reality a measurement can capture is the reality of the measurement itself. So, a physical—that is, empirical—theory should only consider things that are measureable. Physics is what we can see, nothing more and nothing less. Heisenberg hesitates to accept this idea at first, as it reminds him too much of the old philosophical debate about whether a tree makes a noise when it falls over in the forest, if there is no one there to hear it. But now he's ready to explore how far this idea can take his new theory.

The idea simply 'presented itself', Heisenberg would later write. But it presented itself to him and no one else, and he dared to take a step that was no less courageous than that taken by Einstein when he rewrote the seemingly eternal concepts of space and time in his theory of special relativity.

So Heisenberg sets about describing in mathematical terms the transitions from one observable quantum state to another, without referring to the 'in between'. To achieve this, he has to work with a strange kind of mathematical object: number tables. Each table of numbers describes such a transition. Systems develop when such number tables are multiplied with each other. Heisenberg has to puzzle out a way to do this. When two numbers are multiplied, the result is another number—that much is clear. But when two tables of numbers are multiplied, the result is a tangle that needs to be sorted out: which numbers are important, and which can be disregarded? Heisenberg has to learn a whole new way of multiplying, and his feelings vacillate between an awareness of the enormity of his physical insight and the desperation of schoolboy struggling with his difficult maths homework.

However, he soon pulls himself together and formulates a set of conditions: he will only deal with observable quantities, and there will be no violation of the principle of the conservation of energy. A trick allows him to make progress: if you allow two of these transitions between states to occur consecutively, the result must always be a meaningful transition. He continues tinkering with the numbers, miscalculates, corrects himself, and gropes his way forward, getting increasingly excited and increasingly sure in 'the feeling that, through the surface of atomic phenomena, I was looking at a strangely beautiful interior'. When Heisenberg finally manages to note down the multiplications correctly, everything falls into place, as if by magic, to form a new theory of mechanics—quantum mechanics! 'I felt almost giddy,' Heisenberg later writes, 'at the thought that I now had to probe this wealth of mathematical structures nature had so generously spread out before me.'

His hay fever may be gone, but it is now a new, inner fever that keeps him awake at night. Undeterred this time, he gets the final results down on paper one night around 3.00 am.

Too excited to sleep and almost drunk with exhilaration, Heisenberg goes out into the breaking dawn and walks to the southern point of the island to climb the 'Monk'—a fifty-five-metre-high sandstone crag. Arriving safely at the top of the natural tower, Heisenberg watches as the sun rises, and then descends safely again. It's the most precarious rock-climb in the history of physics. Heisenberg's head is the only place where the new theory currently exists. Had he fallen, his theory of quantum mechanics would have been lost forever.

Heisenberg has made his greatest discovery on an island, in isolation, but his colleagues do not leave him alone with it. Once he's back in Göttingen, he discusses his idea with Born, who, to Heisenberg's great surprise, accepts and welcomes it. Heisenberg hastily dashes off a research paper to inform the world of his discovery, sends it to the journal *Zeitschrift für Physik*, and sets off on a short summer trip to Leiden and Cambridge. Although he doesn't mention his breakthrough in the lectures he gives there, he does discuss it in private with his fellow physicists. In his paper, Heisenberg sums up his manifesto for quantum mechanics in one sentence:

> It seems sensible to discard all hope of observing hitherto unobservable quantities, such as the position and period of the electron, and at the same time to admit that the partial agreement of the quantum rules mentioned here with experience are more or less random; instead it seems more reasonable to try to establish a theoretical quantum mechanics, analogous to classical mechanics, but in which only relations between observable quantities occur.

On 19 July 1925, a month after Heisenberg's Heligoland epiphany, Max Born is travelling by train to Hanover to attend a congress of the German Physical Society, when he's overcome by a strange feeling of déjà vu.

Born sits in his train compartment, reading, writing, thinking. He's slept badly, unable to get one nagging thought out of his head: there's something familiar about Heisenberg's idea, but he just can't put his finger on it. Then it dawns on him: what Heisenberg has cobbled together is none other than an outlying branch of mathematics that has until now been deemed useless in practice: matrix algebra. Very few mathematicians, and practically no physicists, are familiar with the concept of matrices. But Born is the exception. He recalls having read about them many years earlier, when he was still considering mathematics as an academic career.

Thus Born recognises that there is already a branch of mathematics that can be useful for quantum mechanics. At some point on the journey, Born's former student Wolfgang Pauli joins him in his compartment, having travelled from Hamburg. Born is excited and keen to explain his insight to Pauli. But Pauli is anything but excited, replying, 'I know you are fond of tedious and complicated formalism. You are only going to spoil Heisenberg's physical ideas with your futile mathematics.'

When Born gets off the train in Hanover, he's depressed and tired from all the thinking he's done. His student Pascual Jordan, a talented mathematician with a severe stammer and restless eyes behind thick glasses, has also been sharing the compartment with him and Pauli, silently following their conversation. Now he speaks up: 'Professor, I know something about matrices—can't I help you?' Born and Jordan set to work, and together they come to recognise the significance of the order in which matrices are multiplied. Two times seven apples is the same as seven times two apples—a property that mathematicians call commutativity. But for matrices, the rule '$a \times b = b \times a$' does not generally apply. This insight is, in fact, included in Heisenberg's initial research paper, but is rather hidden away. Born and Jordan send their follow-up paper to the *Zeitschrift für Physik* just two months after Heisenberg has delivered his. In it, they refine Heisenberg's theory to create what will soon become known as 'matrix mechanics'. It has almost nothing to with the real-world mechanics established by Isaac Newton centuries earlier. 'My miserable efforts,' Heisenberg writes to Pauli, 'are aimed at putting a final end to the concept of orbits that cannot be observed, and replace it with something appropriate.'

Heisenberg returns from Cambridge and his summer vacation with the Boy Scouts. Before the end of the year 1925, Heisenberg, Born, and Jordan author a third research paper, on matrix mechanics, which has gone down in history as the *Dreimännerarbeit* ('paper by three authors'). It further refines and expands matrix mechanics.

Heisenberg's already large ego is bolstered by the fact that his intuition for physics has set him on a promising and previously untrodden path. But he shares the scepticism of his friend Pauli. He dislikes the term 'matrix mechanics', as it smacks too much of the type of pure mathematics that so many physicists find off-putting.

It seems a teacher–student conflict is in the offing. And, indeed, for the rest of his life, Born bears a grudge over the fact that his and Jordan's contributions to quantum mechanics are overlooked or underappreciated. After all, it was in one of his papers that the term 'quantum mechanics' first appeared. Yes, he admits, it was 'terribly clever' of Heisenberg to think of using matrix algebra without having a clue about what matrices are. However, he has trouble admitting that

Heisenberg's bold step was the crucial one. In fact, Heisenberg is not a mathematical dunce who happened to be struck by a bolt of genius. He has something that Born lacks: a profound intuition for physics.

Matrix mechanics doesn't get an enthusiastic reception from physicists. Most scientists now have to learn a whole new type of mathematics and try to understand what these matrices mean for physics. It's extremely complex. The mathematical physicists assure the scientific community that it's all logically sound and able to solve many of the formal puzzles of quantum theory. That's all well and good, but what is it for?

Pauli remains sceptical. Shortly after the publication of Born and Jordan's article, he writes to a colleague: 'Heisenberg's mechanics have restored my vigour and my hope. Although it doesn't solve the puzzle, I believe that it is now once again possible to make progress. First we must try to free Heisenberg's mechanics further from the surge of scholarship in Göttingen and better lay bare its core.'

Heisenberg is not always able to rise above Pauli's gibes. Clearly angry, he writes to Pauli: 'It really is a pigsty that you cannot stop indulging in this slanging match. Your eternal reviling of Copenhagen and Göttingen is a shrieking scandal. When you reproach us that we are such great asses that we have never produced anything new in physics, it may well be true. But then, you are also an equally big jackass because you have not accomplished it either ... (The dots denote a curse of about two minutes' duration!)'

This hits a nerve. Pauli sets to work. In less than a month, he mathematically derives the Balmer series of spectral line emissions of the hydrogen atom, as Bohr had done years earlier with his first model of the atom. With his calculations, Pauli demonstrates that matrix mechanics is not just a mathematical construct, but a useful tool. Heisenberg is appeased. 'I'm sure I don't need to tell you,' he writes to Pauli, 'how happy I am about the new theory of hydrogen and how much I admire the fact that you produced this theory so quickly.'

But Pauli's proof hasn't made matrix mechanics any easier to understand. The majority of physicists are still daunted by the highly complex mathematics. The claim that this is not just a matter of sophisticated mathematics, but also of profound physics, remains an issue of faith for them.

There's another person hoping to make a scientific splash in the summer of 1925. Albert Einstein presents a concept for a 'unified field theory' to the academy in Berlin. Only a week later, he receives a message from Max Born in Göttingen, informing him that Heisenberg has just completed a paper that 'looks very mystical, but is certainly correct and profound'. In a stroke of genius and still only twenty-four years of age, Werner Heisenberg has come up with a new quantum theory—the start of a two-year period of feverish creativity that culminates in a theory that is to shape science in the twentieth century like no other. On reading Heisenberg's paper, Einstein is unimpressed. 'Heisenberg had laid a big quantum egg. In Göttingen they believe it. I don't,' he writes to Paul Ehrenfest on 20 September 1925. And Einstein remains unconvinced for the rest of his life. The man who had been a driving force in promoting the 'old' quantum theory until the first few months of 1925 will play no constructive part in developing the 'new' theory. The year 1925 marks the end of the Einstein era, and the start of Heisenberg era. 'Unified field theory' will never be completed.

In April 1947, the British Navy gathers all the ammunition left over from the war, all the unexploded artillery shells and torpedos, in a concrete cellar on the island of Heligoland: a total of 7,000 tonnes of explosives. The British then attempt to blow up the entire island with the largest non-nuclear explosion in history to that date. The blast has about half the explosive force of the bomb that destroyed Hiroshima, but the island stands firm. Only the sandstone tower that Heisenberg climbed with his head full of quantum mechanics collapses into the sea.

CAMBRIDGE, 1925

The quiet genius

Cambridge, summer 1925. News of Heisenberg's quantum mechanics is spreading among physicists. Many are amazed by the theory; very few understand it. They're wary of matrices, those mathematical monsters. But one taciturn young physicist in Cambridge is not afraid of them. For Paul Dirac—eight months Heisenberg's junior at the age of twenty-three—no maths is too complicated.

Werner Heisenberg visits Cambridge in July 1925, shortly after his epiphany on the island of Heligoland. In Cambridge, he delivers a talk with the enigmatic title 'Term Zoology and Zeeman Botany'. He's still hesitant to speak publicly about his discovery, but he does tell his fellow physicist Ralph Fowler about it. Once back in Göttingen, Heisenberg sends Fowler a copy of his first research paper on quantum mechanics. Fowler passes the paper on to his highly gifted student Dirac, whom Heisenberg didn't meet during his Cambridge visit. Dirac reads the paper, and is one of the few scientists to understand it. Indeed, he picks it up and takes it to another level.

Dirac immediately sees what Born and Jordan had also recognised: the significance in Heisenberg's quantum mechanics of the non-commutative nature of matrices. Working alone, without knowledge of the efforts of the Göttingen team, Dirac completely reinvents the theory, developing his own mathematical formulation of quantum mechanics, similar to that of Born and Jordan, but on a different basis. Rummaging around in an obscure corner of classical mechanics, Dirac stumbles upon a differential operator that, like matrices, satisfies Heisenberg's multiplication rule. Dirac makes use of an elegant mathematical tool developed by the Irish mathematician William Hamilton in the nineteenth century. No more matrices! Or, at least, almost none. Elements similar to matrices appear in his calculations, but only incidentally.

Paul Dirac sends a paper detailing his results to Göttingen, astonishing the luminaries of quantum mechanics there. Dirac? Who's he? Never heard of him. Such a beautiful mathematical structure, such an elegant combination of classical physics and quantum mechanics, had never been seen before. Max Born later described the first time he read Dirac's paper as 'one of the greatest surprises of my life. The author appeared to be very young, but everything was perfect in its way and admirable.'

Heisenberg is also impressed by Dirac's work. Shortly after receiving a proof copy of Dirac's paper, Heisenberg writes a two-page letter in German to Dirac, saying, 'I read your extraordinarily beautiful work on quantum mechanics with great interest. There can be no doubt that all your results are correct, insofar as one believes in the new theory.' Heisenberg even admits to Dirac that his paper is 'actually also considerably better written and more focused than our efforts here'. But then follows disappointment for Dirac: 'I hope you are not disturbed by the fact that part of your results have already been found here some time ago.' Max Born in Göttingen and Wolfgang Pauli in Hamburg have already made some of the discoveries Dirac describes in his paper. However, Dirac deals with any disappointment he may have felt, and Heisenberg's letter marks the start of a long friendship between the two men.

It seems it's all slowly coming together. But it remains confusing. Quantum mechanics can be understood within two different, but obviously related, mathematical systems. Unsurprisingly, the Göttingen set prefers the matrices. The Copenhagen circle surrounding Niels Bohr prefers Dirac's elegant and—as it turns out—more effective version. Beyond these very select circles, physicists view these matrices and operators with increasing despair, asking themselves if there will ever be a way of formulating quantum mechanics that they can also understand.

Paul Dirac is a taciturn man of unremarkable appearance. He is of slight build and medium height, with a wide forehead, a straight, sharp nose, square chin, thin moustache, and clever, pensive eyes. He's the most brilliant of the scientists working on quantum mechanics, but also the strangest. It will later turn out that he has a form of autism.

Born in 1902, Paul Adrien Maurice Dirac described his own childhood as 'unhappy'. This is mainly due to his difficult relationship with his father.

Dirac's taciturn nature stems from that childhood. He has two siblings: a brother and a sister. His Swiss father is extremely strict, demanding unconditional discipline of his children. Dirac's mother is a Cornishwoman. His father insists on speaking only French with the children, while his mother speaks only English. So, every day, the young Paul is pulled in two directions, having to speak only French to his father and only English to his mother and siblings. Paul has to eat in the dining room with his father, speaking French, while the rest of the family eat (and speak English) together in the kitchen. 'Since I found that I couldn't express myself in French, it was better for me to stay silent,' Dirac says later. The family has few friends, and never receives visitors or visits others, so the young Paul has little opportunity to practise and improve his English. His only option is to remain silent, and that is to become the habit of a lifetime for him.

Dirac's siblings go on to become respectable academics, but Paul surpasses them both early. He's the fastest learner, comes top of the class in all scientific and technical subjects, and his abilities in maths and physics soon surpass those of his teachers. Eventually, he even has to set his own homework.

However, the model student has no real idea what he should do with his life, so he follows his brother in studying the subject their father chose for both of them—engineering. When one of Dirac's lecturers witnesses him struggling with practical experiments in the college basement, he advises him: 'Stop messing around here! You're such a talented mathematician. Change courses and do a maths degree! You could finish it in two years.' Paul Dirac takes the advice, and switches to mathematics.

Dirac is now both an engineer and a mathematician, with his feet firmly on the ground and his head in the clouds. He develops both an intuition for connections in physics and an extraordinary sense of the pure beauty of mathematical concepts. Mathematical beauty is something Dirac talks about a lot. He sees beauty in mathematical structures and arguments the way that other people see it in paintings or poetry.

Paul Dirac's epiphany moment comes in 1919, when Einstein's theory of relativity is confirmed by measurements taken during the solar eclipse. And now Dirac knows what he wants to do with his life: theoretical physics. To capture the order of the universe in mathematical equations. He's convinced that nature at the deepest level has a mathematical structure. 'What are the equations?' becomes his usual response to physicists whose bluster about the nature of the world is too vague for his liking.

Dirac moves to Trinity College, Cambridge, in 1921, where Ernest Rutherford is in charge of the Cavendish Laboratory. He's the diametrical opposite of Dirac when it comes to personality: loud, direct, and casually friendly. He has little time for theoreticians, although he accepts Dirac, who visits the lab once a week for tea with the experimenters.

Cambridge is still dominated by the idea that physics has reached full maturity as a science, built upon the two pillars of Newton's classical mechanics and Maxwell's theory of electricity and magnetism. Now there's also this newfangled theory from Germany, Einstein's theory of relativity. Many Cambridge scientists still believe that these theories are all they need to enable them to describe and calculate everything.

Having 'fallen in love'—to use his own words—with Einstein's theory of special relativity, Dirac explores the relationships between the energy, mass, and momentum of a particle. His aim is to describe the very smallest things—atoms—in connection with the fastest speed, which is the subject of the special theory of relativity.

In Cambridge, Dirac remains the shy, reserved person he has always been. Sitting at the high table during one of the traditional formal dinners at Trinity College, one of his fellow diners tries to make polite conversation by asking him about his holiday plans. He makes no reply. Three courses later, he responds, 'Why do you want to know that?' He's not trying to be rude. He simply cannot understand why anyone might be interested in such things. Dirac has absolutely no sense of small talk. He's interested in atoms and the theory of relativity. And his time of greatness is still to come.

LEIDEN, 1925

The prophet of spin

Mid-December 1925. Niels Bohr is in favour; Albert Einstein is against. The disagreement between the greatest minds over quantum mechanics can't go on. Paul Ehrenfest, professor of theoretical physics in Leiden, is friends with both scientists. He invites Bohr and Einstein to his home to try to mediate between them.

When Bohr's train stops in Hamburg on the way to Leiden, Wolfgang Pauli is waiting on the platform. He asks Bohr's opinion of his idea of electron spin. In order to describe the structure of atoms, Pauli has recently ascribed a strange property to electrons, which can only occur in one of two ways, to which he assigns the values ½ and -½. He can't explain what it is; he can only use it to calculate a solution to the puzzle that has stumped theoreticians for years: the splitting of the spectral lines of atoms in a magnetic field, known as the 'anomalous Zeeman effect'.

Two Dutch students at Ehrenfest's institute, Samuel Goudsmit and George Uhlenbeck, have bravely attempted an interpretation of Pauli's discovery. The rotation of electrons around their own axis can be in one of only two directions: to the left or to the right. Goudsmit and Uhlenbeck use the term 'spin' for this rotation. It's a bold interpretation. In physics terms, an electron is a point particle. It has mass, but does not take up any space, and so cannot spin around its own axis. Also, the value ½ means that the electron would be facing back in the original direction after only half a pirouette, so to speak. So, in some ways, a 'spinning electron' can be considered as rotating around its own axis; but in other ways, this idea goes against every concept of classical physics.

Even its creators, Goudsmit and Uhlenbeck, feel so uneasy about this interpretation that they immediately try to suppress it. However, their boss at the institute, Paul Ehrenfest, prevents them from doing so, saying, 'You are both young enough to excuse a folly.' So the image of electrons strangely spinning on their own axis survives. Like most images of quantum phenomena, it is an inadequate stopgap. Spin is a value in the equations of quantum mechanics of which no one has a clear view.

'Very interesting,' says Bohr in answer to Pauli's question on the railway platform. In other words: rejected. Pauli nods. He also rejects Goudsmit and Uhlenbeck's interpretation.

Bohr continues his journey, and is met at the station in Leiden by Albert Einstein and Paul Ehrenfest, who also want to hear Bohr's opinion of the idea of electron spin. Bohr explains his objections. How can an electron moving in the strong electric field of an atom receive enough influence from an external magnetic field to create the finely split spectral lines seen in the anomalous Zeeman effect? Einstein explains to him how this could be possible. He's already guessed Bohr's objection and has dismantled it using his theory of relativity. Bohr is convinced by his argument. He describes Einstein's explanation as a 'revelation'. He now believes the concept of spin will save his atomic model. After all, planets also rotate around their axes as they orbit the sun. This is one final moment of agreement between Bohr and Einstein. They have been unable to agree about quantum mechanics, but they can agree about electron spin.

On the way back to Copenhagen, Bohr's train makes a stop in Göttingen. Werner Heisenberg and Pascual Jordan are on the platform. Electron spin, Bohr informs them, is a great step forward. They believe him.

Bohr travels on to Berlin, to attend a meeting of the German Physical Society marking twenty-five years since the birth of the idea of quanta. Wolfgang Pauli travels from Hamburg. He asks Bohr again about spin, and is not surprised to find Bohr has changed his mind and has now become a prophet of spin. Pauli shakes his head. He considers the 'new gospel' of electron spin to be a false doctrine, and warns Bohr not to spread it further. 'I don't like this thing,' says Pauli. But Bohr does like it. He's now the prophet. And anyone who disagrees with him is a heretic.

AROSA, 1925

A late erotic outburst

December 1925. Erwin Schrödinger, a Viennese physicist now based in Zurich, is spending the Christmas holidays at Villa Herwig, a sanatorium for patients suffering from pulmonary diseases, perched magnificently on the mountainside above the town of Arosa in the Swiss canton of Graubünden. At 1,850 metres above sea level, Arosa's mountain air offers relief to tuberculosis patients. Schrödinger has spent the two previous Christmas holidays here with his wife, Anny. This year, he's accompanied by a woman who is not his wife, but a former girlfriend from Vienna. Anny knows and accepts this, as she has several lovers of her own. They include the mathematician Hermann Weyl, whose wife is the lover of the physicist and friend of Wolfgang Pauli, Paul Scherrer.

Schrödinger is fifteen years older than the youthful adventurers Heisenberg, Pauli, and the rest of the 'physics boys', who are based mainly in Göttingen and Copenhagen. Schrödinger was born in Vienna in 1887 into a wealthy but unconventional family with both English and Austrian roots. He grew up as an only child in the final years of the Austro-Hungarian Empire. Erwin spent his childhood living in an imposing apartment in the centre of Vienna, in the care of several women: his highly sensitive mother and her two sisters, as well as a female cousin and a series of nursemaids. This aspect of Erwin Schrödinger's childhood will shape his relationships with women for his entire adult life. For him, women are creatures who are there to take care of his needs, while their needs are of little consequence to him.

The Schrödingers have little time for music, but they are passionate about the wild, erotic theatre scene of late-nineteenth-century Vienna. At school, Erwin soon becomes known for his extraordinary talents, his self-confidence, his good looks, and his charming, easygoing manner. Schrödinger is a ladies' man. His facial features are sharp, but attractive. However, he is very short-sighted. His eyes seems to float out through the strong, round lenses of his glasses.

In the autumn of 1906, just a couple of weeks after Ludwig Boltzmann's suicide, Schrödinger enrols at the University of Vienna, gaining his doctorate four years later with a thesis entitled 'On the conduction of electricity on the surface of insulators in moist air'. He lands his first paid job as an assistant to the experimental physicist Franz Exner. Schrödinger attends a lecture entitled 'On the current state of the problem of gravitation' given by Albert Einstein in Vienna in 1913, and this sparks in him an interest in the profound, most fundamental questions in physics: the mystery of gravitation, the properties of the cosmos, and the nature of matter.

Then war breaks out. Schrödinger's most important teacher, Fritz Hasenöhrl, the great hope for theoretical physics in Austria, is killed when he's hit in the head by shrapnel from a grenade. Schrödinger also takes part in the war, first as a commissioned officer in the Austrian fortress artillery, and then in the military meteorological service. He hears and sees gunfire, and a medal is pinned to his chest. In 1916, he writes of the war in his diary, 'I think it will continue in its idiotic way, nothing can be done. Frightful. Remarkable: I no longer ask: when will the war be over? But: will it be over? Not naively: hopefully. Are 14 months such a terribly long time? That one already actually begins to doubt the end?'

Once the war is over, Erwin Schrödinger pursues a respectable but unexceptional academic career. He transfers to Jena to work as assistant to Max Wien, then moves on to associate professorships in Stuttgart and Breslau. In 1921, he finally lands a job that pays enough money for him to marry his fiancée, Annemarie 'Anny' Bertel: a full professorship in theoretical physics at the University of Zurich. He's adored and cared for by his wife, but sees no reason to end the constant stream of love affairs with the wives of friends that he's been having since shortly after the couple's marriage, even during their honeymoon. In a few years' time, he'll have fathered three children by three different women, but none with his wife.

But he does provide financially for her, and one of his main aims is to make sure his wife will be financially secure in her old age. When negotiating the pay for his professorship in Zurich, he makes sure his pension entitlements will be transferred to her after his death.

Life in neutral Switzerland is far more pleasant than in post-war Vienna. Schrödinger spends his time debating with the physicist Pieter Debye and the mathematician Hermann Weyl, corresponding with Albert Einstein, Arnold Sommerfeld, and Wilhelm Wien, and carrying out research into the thermodynamics of solid bodies, statistical mechanics, general relativity, atoms and spectra, and the measurement and perception of colour. The research papers he publishes are sound but unspectacular.

Schrödinger is interested in the most hotly discussed modern subjects in physics, but remains stuck in his traditional thinking. He's dubious about the idea that the electrons in Bohr and Sommerfeld's model of the atom can abruptly jump from one orbit to another. He believes there's no place in physics for such changes in state with no interim transition, as they entail things happening with no discernible cause. It's the same objection Einstein has to Bohr's theory.

When the first rumours about Prince de Broglie's interpretation of electron orbits as standing waves begin to circulate, they ring a bell with Schrödinger. The same concept was included in the theoretical results he himself published only a year earlier, albeit very well hidden and previously unrecognised as such by Schrödinger himself.

De Broglie is a theoretical fumbler who had a brilliant idea. Schrödinger is a master of mathematics. He picks up on de Broglie's idea and turns it into a fully fledged theory.

In 1925, as Heisenberg is wrestling with an unfamiliar kind of formalism on the island of Heligoland, Schrödinger pens a report elaborating on de Broglie's electron waves, and even incidentally expressing his suspicion that particles are in reality not particles at all, but rather 'foam crests on a radiating wave that forms the substratum of the world'. He creates an image of the physical world guided by his sense of poetry.

When Schrödinger enthuses about particle waves to one of his academic colleagues in Zurich, the co-worker responds by asking, 'If electrons are waves, what's their wave equation? What medium do the waves move through? What do the waves look like?' Schrödinger has no answer to any of these questions. De Broglie had calculated only one wavelength, and nothing more. So far, his waves are nothing more than an abstract idea, unrooted in the realities of physics. Schrödinger determines to come up with an equation to describe them, which will do away with Bohr's quantum leaps and Heisenberg's cumbersome matrices once and for all. He eventually calculates a wave equation that is eloquent, but wrong. Fit only for the wastepaper basket.

When Schrödinger sets off to spend the Christmas holidays in Arosa in 1925, he's determined to have worked out the right equation by the time he returns home. Maybe his head is cleared by the mountain air, or perhaps he's buoyed by the presence of his mysterious female companion. The maths is hard, and Schrödinger has stupidly forgotten to pack his book on differential equations. 'If only I could do more maths!' he complains. Starting afresh, he simply ignores complications such as relativity, and eventually finds what he's been searching for: a wave equation that formally defines de Broglie's idea. He develops his equation almost unaided, only receiving a little help with a few mathematical details from Hermann Weyl after he returns to Zurich. When Schrödinger's boss asks him whether he enjoyed the skiing in Arosa, he answers that he was distracted by 'a couple of calculations'.

Schrödinger's equation is an elegant mathematical construct. It describes a field that is governed by a kind of energy function represented by a mathematical operator. When Schrödinger applies it to an atom, it results in a number of solutions, each of which describes a static pattern in the field. These are the energy states of the atom in question. With a skilful mathematical manoeuvre, Schrödinger has conjured up the quantum rules that Bohr invoked so crudely. His equation means that the energy states of an atom are now no more mysterious than the harmonic tones produced by a vibrating violin string.

Schrödinger transforms quantum leaps—those sudden, discontinuous changes in states considered so unacceptable by some physicists—into flowing transitions from one standing wave pattern to another. He has transformed atomic physics from a patchwork into an artwork. Schrödinger's equation is not derived through plain logic; it is composed like a piece of music.

As early as January the next year, Schrödinger writes his first paper presenting his equation, and submits it to the journal *Annals of Physics*. This marks the start of an extraordinarily creative time for Schrödinger. He produces four more papers before June—almost one per month—each further developing his theory of wave mechanics.

The old guard is delighted. Arnold Sommerfeld says Schrödinger's treatise struck him 'like a thunderbolt'. 'The idea behind your article shows true genius!' Albert Einstein writes to Schrödinger, following up a little later with, 'I am convinced that you have made a decisive advance with your formulation of the quantum condition, just as I am equally convinced that the Heisenberg-Born route is off the track.' The highly gifted American physicist J. Robert Oppenheimer, currently in Göttingen studying quantum physics under Max Born, enthuses, 'Here's this very pretty theory, possibly one of the most perfect, accurate, and beautiful theories ever developed by humankind.'

Bohr and Heisenberg's circle of quantum researchers also sit up and take notice. 'I believe this paper is among the most important to have been written in recent times,' writes Wolfgang Pauli in Hamburg to Pascual Jordan in Göttingen. 'Read it carefully and with reverence.'

It seems the classical order has been restored. Schrödinger's equation becomes the cornerstone of this new quantum physics. The man whose name it bears, now almost forty years of age, has produced his 'greatest work during a late erotic outburst in his life', as observed by his good friend Hermann Weyl, the mathematician who is also Schrödinger's wife's lover. It's one of many erotic outbursts so far in Schrödinger's life, and it won't be the last. But this is the one that leads to his most important discovery. Both Schrödingers maintain a discreet silence over the identity of the 'Dark Lady' who accompanied Erwin to Arosa that year. In the world of physics, Schrödinger is a traditionalist. In real life, he is anything but.

Sometime during the year 1926, Schrödinger begins giving private maths lessons to two girls called Itha and Roswitha Junger, affectionately known as Ithi and Withi. They are the fourteen-year-old daughters of an acquaintance of Schrödinger's wife, Anny—twins, but not identical. They attend a convent school, where they are in danger of failing the year. Schrödinger saves them from failing, explains his research to them, talks to them about religion, writes terrible poetry to them, and there is also 'a great deal of caressing and cuddling', as Itha later recounts. The girls are flattered by his attention. Schrödinger falls in love with Itha. He waits until she has turned sixteen, and then, during a ski trip, he comes to her room in the middle of the night and confesses his love for her. Shortly after her seventeenth birthday, they begin a love affair.

COPENHAGEN, 1926

Waves and particles

Copenhagen, 1926. Heisenberg moves into a cosy little attic flat in the Bohr Institute. It has slanting walls and a view of the Fælledparken gardens. Bohr and his family now live in a grand villa next to the institute. Heisenberg visits them so often that he feels 'half at home' in their house.

Bohr is exhausted. The institute building has been renovated and extended, and the process has sapped Bohr's energy. He's struck by a bad bout of the flu, and spends a full two months recovering. Heisenberg makes use of the time and Bohr's absence to work on the problem of the spectral lines of helium by applying his theory to it. It's an important test of his idea. Both the theory and its creator withstand the test with ease.

As soon as Bohr has recovered, the same old game begins again. 'After eight or nine in the evening, Bohr would enter my room and ask, "Heisenberg, what do you think of this problem?" Then we would talk and talk, often continuing until midnight or one in the morning.' Sometimes Bohr also invites Heisenberg to the villa for an engaging chat that lasts long into the evening and is accompanied by many a glass of wine.

At the same time, Heisenberg has to honour his teaching duties. He gives two university lectures a week on theoretical physics—in Danish. At twenty-four, he's barely older than his audience. After attending his first lecture, one student remarks, 'It's hard to believe that he's supposed to be so intelligent,' adding that he looks 'like a smart carpenter's apprentice'. Heisenberg soon grows accustomed to the rhythm of work at the institute. After making friends with colleagues, he goes out sailing, horse riding, and hiking with them on weekends. But, following a visit from Schrödinger in October 1926, he finds he has less and less time for such leisure activities, although they are so important to him. Bohr keeps him on a tight leash. The master needs the attention of his apprentice.

But Bohr and Heisenberg can't agree on an interpretation of quantum mechanics. In his search for clarity, Bohr wants to get to the very bottom of things: at what point on the journey down to the quantum scale does visualisability end? And why does it end? Bohr wants to use the theory to reconcile the wave-particle duality, and thus rescue the concepts of old physics and carry them over into a new era.

Heisenberg can't understand Bohr's concerns. What's the problem? We have a theory. We can make predictions with it and test them experimentally. The new definitions of position and velocity are provided by the theory. Any other ideas we have about it are immaterial.

The pain caused to physicists by the wave-particle duality is almost physical itself. In a letter to Paul Ehrenfest dated August 1926, Albert Einstein writes, 'Here waves, here quanta! The reality of both is rock solid. Only the devil can make any real rhyme or reason out of it.'

In classical physics, all was in order. There were waves and there were particles—and nothing could be both at the same time. But in quantum mechanics, particles sometimes pretend to be waves. Or is it the other way round? Heisenberg's concept of quantum mechanics is based on particles. Schrödinger conceives of the world as being a great tangle of waves. Then the two approaches turn out to be equivalent from a mathematical point of view. How is that possible? Two fundamentally different, seemingly irreconcilable, approaches that lead to the same result? The proof of the equivalence of matrix mechanics and wave mechanics has not helped resolve the particle-wave duality problem.

It almost seems as if the electrons are deliberately trying to make fools of those who study them: as long as no one is watching, they're waves. As soon as someone looks, they're suddenly particles. What is the mechanism behind this, what's the cause and what's the effect? The more Bohr and Heisenberg think about it, the less they understand it. Together they try to lay bare the core of this paradox—but only succeed in laying bare the growing tensions between them. They have two irreconcilable approaches, but neither is making any progress. Each loses patience with the other, and they end up almost driving each other crazy.

Heisenberg believes the answer lies in the theory itself: what can it tell us about the nature of reality at the atomic scale? And if the answer involves quantum leaps and discontinuities, then so be it. For him, it's already clear that the particle side is the dominant one in the wave-particle duality. He refuses to give space to anything reminiscent of Schrödinger's waves. Bohr is more open. Unlike Heisenberg, he's not fixated on matrix mechanics. He wants to play with both concepts: particles and waves. Mathematical formalisms do not enchant Bohr. Heisenberg's thinking always starts with maths. Bohr is trying to fathom physics beyond the mathematical. He's searching for a physical interpretation of the wave-particle duality. Heisenberg is seeking a way to describe it with mathematics.

Bohr is searching for a way to accommodate both waves and particles in a complete description of atomic processes. He wants to reconcile those two contradictory concepts. Bohr believes achieving this will be the key that opens the door to an understanding of quantum mechanics.

BERLIN, 1926

A visit with the demigods

Berlin, April 1926. From their vantage point at the summit of the Mount Olympus of physics, Albert Einstein and Max Planck watch these scenes unfold. Einstein is now approaching fifty, and Planck is almost seventy years of age. Both physicists are considered demigods of science. They can summon the young heroes of physics at their pleasure, to receive reports of their quantum quests. Twenty-four-year-old Werner Heisenberg now has the honour of visiting the 'stronghold of physics', as he likes calls Berlin, to give an account of his latest feats. Heisenberg still looks like the boyish lad who went hiking with Bohr in the Hainberg Hills four years earlier, and who wavered between timidity and brashness. He's still wrestling with those strange quanta. But now he's less of a waverer. After all, it was he who discovered quantum mechanics. Not Planck, nor Einstein. The theory is his creation.

Einstein has now become a cult figure. He's immediately recognisable, with his wild white hair, bright eyes, and always-slightly-too-short trousers. He's a cultural icon. As his social status changes, so does his way of thinking and talking about science. Gods proclaim the truth, and don't give reasons. 'Quantum mechanics is very awe-inspiring,' he writes to Max Born, 'but an inner voice tells me that it is not yet the real thing. The theory says a lot, but does not really bring us any closer to the secret of the Old One. I, at any rate, am convinced that He does not play dice.'

Now, standing in front of the chalkboard on 28 April 1926, his notes spread out before him, the twenty-four-year-old Werner

Heisenberg has every reason to feel nervous. He's been asked to explain his matrix mechanics to the venerable physics colloquium at the University of Berlin. Munich and Göttingen were mere rehearsals for this. Berlin is the real premiere. Heisenberg scans the audience. Max von Laue, Walther Nernst, Max Planck, and Albert Einstein are all at the front of the crowd—four Nobel Prize winners sitting in a row.

Heisenberg has often heard these men's names, and has often uttered them himself. Now he's face-to-face with them for the first time. He's aware that they're no longer young. That their great feats of physics were achieved quite some time ago. As he later recalls, he 'went to great lengths to present as clearly as possible the concepts and mathematical foundations of the new theory that were so unfamiliar in the physics of the time'.

Einstein's interest is sufficiently aroused for him to grant Heisenberg an audience to talk more about it. As the colloquium disperses after the lecture, Einstein approaches Heisenberg and invites him to his home for a visit. They walk together through the city streets towards Einstein's apartment. This stroll with a demigod, rather than the lecture itself, will stick in Heisenberg's memory of his trip to Berlin. At last, he has managed to meet Einstein, in the flesh! Heisenberg had hoped to meet Einstein during a physicists' congress in Leipzig four years earlier, but Einstein had chosen to stay home due to the murder of the foreign minister, Walther Rathenau, while Heisenberg was unable to travel as he had recently been robbed and could not afford the fare.

Now, as they stroll through Berlin on this spring day, Einstein takes charge of the conversation. He asks about Heisenberg's family, his education, and his past research. He doesn't yet mention Heisenberg's new theory at all. It's not until they arrive at Einstein's apartment on the fourth floor of the building at Haberlandstrasse 5 that the host moves the conversation on to more weighty topics. We owe the following account to Heisenberg's memoir, *Physics and Beyond*.

Einstein dislikes the fact that Heisenberg's matrix mechanics banishes position and velocity from the central focus of physics, replacing them with abstruse mathematical constructs. 'What you have told us sounds very strange,' he says to Heisenberg, challenging him. 'You assume the existence of electrons inside the atom, and you are probably quite right to do so. But you refuse to consider their orbits, even though we can observe electron tracks in a cloud chamber. I should very much like to hear more about your reasons for making such strange assumptions.'

This is the chance Heisenberg has been hoping for—an opportunity to get the forty-seven-year-old doyen of quantum physics on his side. He's only done what reality forced him to do, Heisenberg tells him. He doesn't want a theory built on unknown and possibly unknowable quantities; he wants a theory built on the actual observations that scientists make of atoms. 'We cannot observe electron orbits inside the atom,' he explains, 'but the radiation that an atom emits during discharges enables us to deduce the frequencies and corresponding amplitudes of its electrons. After all, even in the older physics, wave numbers and amplitudes could be considered substitutes for electron orbits. Now, since a good theory must be based on directly observable magnitudes, I thought it more fitting to restrict myself to these, treating them, as it were, as representatives of the electron orbits.'

No, Einstein doesn't find this 'fitting' at all. 'But you don't seriously believe,' he counters, 'that none but observable magnitudes must go into a physical theory?'

Yes, that's exactly what Heisenberg does believe. He tries to turn Einstein's own argument against him with a counter-question. 'But isn't that,' he replies, 'precisely what you took as the basis for your theory of relativity? After all, you did stress the fact that it is impermissible to speak of absolute time, simply because absolute time cannot be observed; that only clock readings, be it in the moving reference system or the system at rest, are relevant to the determination of time.'

This hits the mark with Einstein. 'Possibly I did use this kind of reasoning,' he mumbles in reply, 'but it is nonsense all the same. Perhaps I could put it more diplomatically by saying that it may be heuristically useful to keep in mind what one has actually observed.

But in principle, it is quite wrong to try founding a theory on observable magnitudes alone. In reality, the very opposite happens. It is the theory that decides what we can observe. You must appreciate that observation is a very complicated process. The phenomenon under observation produces certain events in our measuring apparatus. As a result, further processes take place in the apparatus, which eventually and by complicated paths produce sense impressions and help us to fix the effects in our consciousness. Along this whole path—from the phenomenon to its fixation in our consciousness—we must be able to tell how nature functions, must know the natural laws at least in practical terms, before we can claim to have observed anything at all. Only theory, that is, knowledge of natural laws, enables us to deduce the underlying phenomena from our sense impressions.

'When we claim that we can observe something new, we ought really to be saying that, although we are about to formulate new natural laws that do not agree with the old ones, we nevertheless assume that the existing laws—covering the whole path from the phenomenon to our consciousness—function in such a way that we can rely on them and hence speak of "observations". In the theory of relativity, for instance, we presume that, even in the moving reference system, the light rays moving from the clock to the observer's eye behave more or less as we have always expected them to behave. And in your theory, you quite obviously assume that the whole mechanism of light transmission from the vibrating atom to the spectroscope or the eye works just as one has always supposed it does, essentially according to Maxwell's laws. If that were no longer the case, you could not possibly observe any of the magnitudes you call observable. Your claim that you are introducing none but observable magnitudes is therefore an assumption about a property of the theory that you are trying to formulate. You are, in fact, assuming that your theory does not clash with the old description of radiation phenomena in the essential points. You may well be right, of course, but you cannot be certain.'

Heisenberg is taken aback by this answer. He had thought Einstein was on his side when it came to the question of observability. Einstein's arguments make sense to Heisenberg. But aren't they equally an argument against Einstein's own theory of relativity? Is Einstein so keen to attack quantum mechanics that he's prepared to topple his own masterpiece? To back up his arguments, Heisenberg draws on one of Einstein's old allies: the philosopher Ernst Mach. 'The idea that a good theory is no more than a condensation of observations in accordance with the principle of thought economy surely goes back to the physicist and philosopher Mach, and it has, in fact, often been said that your relativity theory makes decisive use of Machian concepts. But what you have just told me seems to indicate the very opposite. What am I to make of all this, or rather what do you yourself think about it?'

Einstein had indeed studied the work of the Austrian philosopher Ernst Mach, back when he was still a patent officer in Bern. Mach had redefined the goal of science as being not to unravel the essence of nature, but to find the simplest and most economical expression of facts—that is, of empirical data. Every scientific concept is defined by a description of how it must be measured. It was on the basis of this Machian philosophy of science that Einstein undertook to dismantle the old concepts of absolute time and space. But Heisenberg clearly does not know that Einstein later discarded Mach's philosophy because it neglects to take into account the fact that the world actually exists in reality.

'It's a very long story,' Einstein says, 'but we can go into it if you like. Mach's concept of thought economy probably contains part of the truth, but strikes me as being just a bit too trivial. Let me, first of all, produce a few arguments in its favour. We obviously grasp the world by way of our senses. Even when small children learn to speak and think, they do so by recognising the possibility of describing highly complicated but somehow related sense impressions with a single word—for instance, the word "ball". They learn it from adults, and get the satisfaction of being able to make themselves understood. In other words, we may argue that the formation of the word, and hence of the concept, "ball" is a kind of thought economy enabling the child to combine very complicated sense impressions in a simple way. Here Mach does not even enter into the question of which mental and physical predispositions must be satisfied in man—or, in this case, the small child—before the process of communication can be initiated. With animals, this process works considerably less effectively, as everyone knows, but we shan't talk about that now.

'Now, Mach also thinks that the formation of scientific theories, however complex, takes place in a similar way. We try to order the phenomena, to reduce them to a simple form, until we can understand what may be a large number of them, whereby "understand" means "be able to grasp them in all their diversity" with the aid of a few simple concepts. All this sounds very reasonable, but we must nevertheless ask ourselves in what sense the principle of mental economy is being applied here. Are we thinking of psychological or of logical economy, or, again, are we dealing with the subjective or the objective side of the phenomena? When the child forms the concept of "ball", does he introduce a purely psychological simplification, in that he combines complicated sense impressions by means of this concept, or does this ball really exist? Mach would probably answer that the two statements express one and the same fact. But he would be quite wrong to do so. To begin with, the assertion "the ball really exists" also contains a number of statements about possible sense impressions that may occur in the future. Now future possibilities and expectations make up a very important part of our reality, and must not be simply forgotten. Moreover, we ought to remember that inferring concepts and things from sense impressions is one of the basic presuppositions of all our thought. Hence, if we wanted to speak of nothing but sense impressions, we should have to rid ourselves of our language and thought. In other words, Mach rather neglects the fact that the world really exists, that our sense impressions are based on something objective.

'I have no wish to appear as an advocate of a naïve form of realism; I know that these are very difficult questions, but then I consider Mach's concept of observation also to be far too naïve. He pretends that we know perfectly well what the word "observe" means, and thinks this exempts him from having to discriminate between "objective" and "subjective" phenomena. No wonder his principle has so suspiciously commercial a name: "thought economy". His idea of simplicity is much too subjective for me. In reality, the simplicity of natural laws is an objective fact as well, the correct conceptual scheme must balance the subjective side of this simplicity with the objective. But that is a very difficult task.

'Let us rather return to your lecture. I have a strong suspicion that, precisely because of the problems we have just been discussing, your theory will one day get you into hot water. I should like to explain this in greater detail. When it comes to observation, you behave as if everything can be left as it was—that is, as if you could use the old descriptive language when talking about what physicists observe. In that case, however, you will have to say: in a cloud chamber we can observe the path of the electrons. At the same time, you claim that there are no electron paths inside the atom. This is obvious nonsense, for you cannot possibly get rid of the path simply by restricting the place in which the electron moves.'

Now it's up to Heisenberg to defend his theory. He attempts to do so by allowing some of Einstein's objections. 'For the time being, we have no idea in what language we must speak about the processes inside the atom. True, we have a mathematical language, that is, a mathematical scheme for determining the stationary states of the atom or the transition probabilities from one state to another, but we do not know—at least, not in general—how this language is related to normal language. And, of course, we need this connection if we are to apply this theory to experiments in the first place. For when it comes to experiments, we invariably speak in the traditional language of classical physics. Hence, I cannot really claim that we have "understood" quantum mechanics. I assume that the mathematical scheme works, but no link with the traditional language has been established so far. And until that has been done, we cannot hope to speak of the path of the electron in the cloud chamber without inner contradictions. Hence it is probably much too early to solve the difficulties you have mentioned.'

'Very well, I will accept that,' Einstein relents. 'We shall talk about it again in a few years' time. But perhaps I may put another question to you. Quantum theory as you have expounded it in your lecture has two distinct faces. On the one hand, as Bohr has rightly stressed, it explains the stability of the atom; it causes the same forms to reappear time and again. On the other hand, it explains that strange discontinuity or inconsistency of nature that we observe quite clearly when we watch flashes of light from a radioactive compound on a scintillation screen in the dark. These two aspects are obviously connected. In your quantum mechanics you will have to take both into account, for instance when you speak of the emission of light by atoms. You can calculate the discrete energy values of the stationary states. Your theory can thus account for the stability of certain forms that cannot merge continuously into one another, but must differ by finite amounts and seem capable of permanent re-formation. But what happens during the emission of light? As you know, I suggested that, when an atom drops suddenly from one stationary energy value to the next, it emits the energy difference as an energy packet, a so-called light quantum. In that case, we have a particularly clear example of discontinuity. Do you think that this conception is correct? Or can you describe the transition from one state to another in a more precise way?'

Now Heisenberg has to hide behind Bohr. 'Bohr has taught me that one cannot describe this process by means of the traditional concepts—that is, as a process in time and space. With that, of course, we have said very little, no more, in fact, than that we do not know. Whether or not I believe in light quanta, I cannot say at this stage. Radiation quite obviously involves the discontinuous elements to which you refer as light quanta. On the other hand, there is a continuous element, which appears, for instance, in interference phenomena, and which is most simply described by the wave theory of light.

'But you are of course quite right to ask whether quantum mechanics has anything new to say on these terribly difficult problems. I believe that we may at least hope that it will one day. I could, for instance, imagine that we should obtain an interesting answer if we considered the energy fluctuations of an atom during reactions with other atoms or with the radiation field. One could then ask about the fluctuation of the energy within the atom. If the energy should change discontinuously, as we expect from your theory of light quanta, then the fluctuation, or, in more precise mathematical terms, the mean square fluctuation, would be greater than if the energy changed continuously. I am inclined to believe that quantum mechanics would lead to the greater value, and so establish the discontinuity. On the other hand, the continuous element, which appears in interference experiments, must also be taken into account. Perhaps one must imagine the transitions from one stationary state to the next as so many fade-outs in a film. The change is not sudden—one picture gradually fades while the next comes into focus so that, for a time, both pictures become confused, and one does not know which is which. Similarly, there may well be an intermediate state in which we cannot tell whether an atom is in the upper or the lower state.'

'You are moving on very thin ice,' Einstein now warns Heisenberg, 'for you are suddenly speaking about what we know about nature and no longer about what nature really does. In science we ought to be concerned solely with what nature does. It may very well be that you and I know quite different things about nature. But who would be interested in that? Perhaps you and I alone. To everyone else it is a matter of complete indifference. In other words, if your theory is right, you will have to tell me sooner or later what the atom does when it passes from one stationary state to the next.'

Heisenberg hopes he will be able to tell him that one day, and he hopes that Einstein will be satisfied with his answers for now. 'Perhaps,' he replies, 'but it seems to me that you are using language a little too strictly. Still, I do admit that everything that I might now say may sound like a cheap excuse. So let's wait and see how atomic theory develops.'

However, Einstein is not satisfied with this, and continues to push, 'How can you really have so much faith in your theory, when so many crucial problems remain completely unsolved?'

Heisenberg has difficulty formulating an answer. 'Why' questions are not his strong suit. He stops and ponders awhile before replying, 'I believe, just like you, that the simplicity of natural laws has an objective character, that it is not just the result of thought economy. If nature leads us to mathematical forms of great simplicity and beauty—by forms, I am referring to coherent systems of hypotheses, axioms, etcetera—to forms that no one has previously encountered, we cannot help thinking that they are "true", that they reveal a genuine feature of nature. It may be that these forms also cover our subjective relationship to nature, that they reflect elements of our own thought economy. But the mere fact that we could never have arrived at these forms by ourselves, that they were revealed to us by nature, suggests strongly that they must be part of reality itself, not just of our thoughts about reality. You may object that by speaking of simplicity and beauty I am introducing aesthetic criteria of truth, and I frankly admit that I am strongly attracted by the simplicity and beauty of the mathematical schemes with which nature presents us. You must have felt this, too: the almost frightening simplicity and wholeness of the relationships which nature suddenly spreads out before us and for which none of us was in the least prepared. And this feeling is something completely different from the joy we feel when we have done a set task (be it in physics or elsewhere) particularly well. That is one reason why I hope that the problems we have been discussing will be solved in one way or another. In the present case, the simplicity of the mathematical scheme has the further consequence that it ought to be possible to think up many experiments whose results can be predicted very precisely from the theory. And if the actual experiments should bear out the predictions, there is little doubt but that the theory reflects nature accurately in this particular realm.'

That sounds more like a plea than an argument. But Einstein is somewhat propitiated. 'Control by experiment,' he replies, 'is, of course, the trivial prerequisite for the validity of any theory. But one can't possibly test everything. That is why I am so interested in your remarks about simplicity. Still, I should never claim that I really understood what is meant by the simplicity of natural laws.'

Einstein is convinced that the outside world really exists and that human imagination is capable of sounding out its depths. Heisenberg does not trust the imagination of anything beyond our day-to-day world. For him, the numbers must add up; the equations must be correct. Only then can we talk about imagination.

Their conversation has now brought them to the consideration of some deep philosophical questions. What is scientific truth? Is beauty one of its criteria? The two men talk on for a while, before Heisenberg changes the subject. He's at an important juncture in his life. In three days' time, he's due to return to work in Copenhagen, to his double job as both Bohr's assistant and a university lecturer. Bohr would like to keep him in Copenhagen. But the University of Leipzig has just offered him a professorship, a permanent and prestigious position—and an unusual honour for so young a scientist. Where should he go, Copenhagen or Leipzig? Heisenberg asks Einstein's advice. Go back to Bohr, is Einstein's answer.

Heisenberg takes his leave. He's disappointed not to have convinced Einstein. The two will not meet again until a year-and-a-half later, when they will once again argue over quantum mechanics and reality—but then, the gloves will be off.

The next day, Heisenberg informs his parents in Munich that he has decided not to take up the offer from Leipzig. There will be more offers in the future, he tells himself. And if not, then it will be because he does not deserve them.

Shortly after his meeting with Heisenberg, Einstein writes to Arnold Sommerfeld: 'Of the recent attempts at a deeper formulation of quantum laws, I like Schrödinger's the best. I can't help but admire the Heisenberg-Dirac theories, but for me they do not have the whiff of reality.'

BERLIN, 1926

The Plancks throw a party

Berlin, summer 1926. Werner Heisenberg's visit with Einstein and Planck, the demigods of physics, was disappointing for both sides. The demigods favour that other hero, Erwin Schrödinger, who is currently publishing papers like there's no tomorrow. Einstein refers Planck to Schrödinger's papers 'with justified enthusiasm'. 'Not such an infernal machine,' writes Einstein in an oblique reference to Heisenberg's matrix algebra, 'but rather a clear idea and constrained in its implementation.'

There is something tangible about the idea of electrons in atoms being standing waves—a quality that Heisenberg's matrices lack. How can it be that two completely different theories describe the same phenomena? The confusion is cleared up surprisingly quickly—by Schrödinger himself. In spring 1926, he discovered that wave mechanics and matrix mechanics are not all that different after all. Behind their apparently opposing facades lies the same theory, just clad in different mathematics. Schrödinger's waves can be used to calculate numbers that obey the rules of matrix algebra. Matrix algebra, with some tweaking, produces Schrödinger's equation. Schrödinger was not the first to notice this astonishing equivalence between the two theories. Wolfgang Pauli had also discovered it, but did not publish anything about it, only sketching out his proof in a letter to Pascual Jordan.

So now, quantum physicists are faced with a choice between Schrödinger's waves or Heisenberg's matrices. Many physicists incline towards the waves, being familiar with them from traditional physics.

Matrix mechanics is still strange and inscrutable to them. The question remains: is there just one 'correct' way of talking about the smallest building blocks of nature, or is it a question of taste, or of convenience?

It becomes time to invite Schrödinger to Berlin, since he's already touring Germany, to publicise his equation. Schrödinger accepts an invitation from Planck to come to the German capital in July. After spending a few days in Stuttgart, he travels to Berlin, and then on to Jena, where he worked as an assistant five years earlier.

The Plancks meet Schrödinger at the station and take him back to their sober and austerely furnished house at Wangenheimstrasse 21 in the leafy Berlin suburb of Grunewald. The visitor has a full schedule. On 16 July, he gives a talk at the German Physical Society on 'The fundamentals of atomism on the basis of wave theory'. The next day, he gives another, more specialised, lecture to a smaller audience at the university's physics colloquium. All the doyens of physics in Berlin are present—Einstein, Planck, von Laue, and Nernst—and they lend Schrödinger a benevolent ear. At last, here's someone advancing quantum physics in the good old way, with classical concepts and tried-and-tested mathematics. After the lecture, the Plancks throw a party for Schrödinger. Max Planck, fired up by the lecture and tipsy from the party drinks, considers naming Erwin Schrödinger as his successor when he retires in the coming year.

This is the first time Einstein and Schrödinger have had an opportunity to engage in a discussion of any depth. Finally! Schrödinger has long felt a bond with Einstein, ever since he attended that lecture in Vienna in 1913, when Einstein opened the young Schrödinger's mind to the great mysteries of physics. Since then, the two have corresponded by letter, sending each other their latest works. The two men had even met briefly in 1924 at a congress of natural scientists in Innsbruck. Schrödinger never misses an opportunity to stress how great a role model Einstein is for him.

With his difficult talk with Heisenberg still fresh in his memory, Einstein is relieved to find he gets on much better with Schrödinger. He likes him. Schrödinger lacks the north German formality and reserve that Einstein dislikes in Heisenberg, who, despite having been born in Bavaria, is still considered a 'Prussian' or a 'newcomer' in southern Germany due to his Westphalian father's influence on his manner, and even his way of speaking German. Schrödinger is educated and cultivated, approachable and well-mannered, and, with his warm Viennese dialect, a perfect partner for a party conversation.

And Einstein and Schrödinger agree on matters of quantum mechanics. Both believe wave mechanics is the better, the more correct, the true, theory. But the two have more than their scientific theories in common. Both are married, and value the care being in a family provides, but both also seek their pleasures outside their marriage.

Despite those affinities, Einstein does not hesitate to point out the weaknesses he sees in Schrödinger's physics. In his lecture to the German Physical Society, Schrödinger had reiterated his hope that the waves his equation describes would turn out to be the true representation of electrons and other things — not particles, as others imagine them, but wavelike concentrations of mass and charge. Einstein likes this, but remains cautious. Schrödinger has created a cause for hope, but his argument is not totally convincing yet. It might be nothing but wishful thinking.

Werner Heisenberg is far from impressed by Schrödinger's theory, and he doesn't hesitate to express that opinion. 'The more I ponder on the physical part of Schrödinger's theory, the more horrible I find it,' Heisenberg writes to Wolfgang Pauli, 'but please excuse my heresy and don't tell anyone about it.'

Einstein would not be so blunt to Schrödinger to his face, and hopes that he's right and that Heisenberg is wrong, but he senses that this matter will not be as easy to solve as Schrödinger thinks.

GÖTTINGEN, 1926

The abolition of reality

Spring 1926. Max Born has spent five months in the United States. He's experienced the thundering power of the Niagara Falls and the Grand Canyon: 'Suddenly this chasm, this mad plunge, jagged and ruptured, which no imagination could envision.' He's been ice sailing on Lake George, visited the bleak working-class districts of 'the steel city' Chicago, travelled more than 10,000 kilometres by railroad, lectured at top universities, and earned hard dollars. Several universities offer him permanent posts, but Born rejects them, feeling he has an obligation to his homeland. The University of Göttingen makes this decision easier for him by offering to raise his salary. In April, Born arrives back in Germany—and in a difficult situation. Someone had scrawled 'Jew!' in the margin of the letter sent by his boss recommending his pay rise. His assistant, Heisenberg, has left for Copenhagen. Their years of fruitful cooperation are over.

Something that captures Born's attention on his return: the extraordinary series of publications on wave mechanics that Schrödinger is currently pumping out. Born reads them with 'complete surprise'. Even a few brief glances at the papers are enough for the mathematical high-flyer to recognise 'the fascinating power and elegance' of the theory Schrödinger has devised, as well as the 'superior nature of wave mechanics as a mathematical tool'. Even Heisenberg, the creator of matrix mechanics, requires the help of the maths genius Wolfgang Pauli to describe a hydrogen atom—the simplest of all atoms—using matrix mechanics. To do the same with wave mechanics is child's play.

Born is both surprised and a little annoyed. He could have thought of this himself. Albert Einstein had called his attention to Louis de Broglie's bold ideas about matter waves as early as 1924. Born considered it a good theory at the time. 'It could be of great significance,' he wrote to Einstein. After pondering a little over de Broglie's waves, he played around with the equations. But when Born's mathematical skills were needed to tame Heisenberg's monstrous matrices, he put de Broglie's work to one side. Max Born has been a Göttingen man for a long time, and in Göttingen they reckon with matrices.

Born may have been in Göttingen a long time, but he isn't a native. He grew up in Breslau, capital of the Prussian province of Silesia. His first love is mathematics, rather than physics. When he enrols at the University of Breslau, his father, embryology professor Gustav Born, advises him not to specialise too early. Accepting his father's advice, he attends lectures on chemistry, zoology, law, philosophy, and logic, before transferring to mathematics, physics, and astronomy. He also later studies in Heidelberg and Zurich, before finally earning his doctorate in mathematics from Göttingen University in 1906.

As soon as he's gained his doctorate, he begins a year of military service, which is cut short due to his asthma. After six months as a post-doctoral student in Cambridge under J.J. Thomson, he returns to Breslau to begin experimenting in the laboratory. But he soon realises that experimental physics is not his vocation—he lacks the necessary dexterity and patience. He's not a good, or even vaguely competent, experimental physicist.

So Born decides to concentrate on theory, and takes a position as a lecturer in Göttingen University's world-renowned faculty of mathematics, where David Hilbert is in charge. 'Physics is much too hard for physicists,' Hilbert famously opines, believing it should be left to mathematicians. Born accepts a position as professor of theoretical physics in Berlin, shortly before the outbreak of the First World War. Conscripted into the military, Born works first as a radio operator in the air force, and then as an artillery developer. Stationed near Berlin, he's able to attend seminars at the university and spend time playing music with Albert Einstein.

In the spring of 1919, with the war now over, Max von Laue

suggests to Born that they swap jobs. Von Laue is a professor in Frankfurt who won the Nobel Prize for Physics in 1914 for his discovery of the diffraction of X-rays by crystals, which was proof that X-rays behave like waves. He has an 'old and ardent wish' to work with Max Planck, his teacher and idol, in Berlin. 'Be sure to accept,' Albert Einstein advises him, and Born does so. Not two years later, he takes the next step in his career and moves to Göttingen to become head of the Institute for Physics. The institute is little more than a small room, with an assistant and a part-time secretary for Born. He's determined to expand it until it can compete with Sommerfeld's institute in Munich. After wooing them for a long time, Born manages to entice Wolfgang Pauli and Werner Heisenberg to Göttingen, giving an official home to matrix mechanics, the first theory of post-classical physics.

Now, after the publication of Schrödinger's many papers, Born turns his attention back to wave theory. He recognises how powerful the concept is, but he's less dogmatic about it than Schrödinger, whose blanket rejection of particles and quantum leaps is too extreme for him. During his initial studies in Göttingen, Born learned how helpful the particle theory could be in understanding collisions between atoms. Those collisions are between particles, not waves. They have a specific location, rather than being spread out through space like ripples in a pond. Schrödinger's waves simply don't fit into the world that experimental atomic physicists see.

Born uses wave mechanics to calculate what happens when two particles collide, and is astonished by the result. The waves created by the rebounding particles do spread out through space like ripples on the surface of a pond. According to Schrödinger's interpretation, that means that the particles themselves are smeared in all directions. What could that mean? Colliding particles are still particles, even if they are based on wave motion, and not wisps of mist—even Schrödinger believes that. They must be somewhere, in a specific position, and cannot dissipate through space in some mysterious way.

Schrödinger's offers a rather dubious argument to support his interpretation. He demonstrates that the wave packets—which is the term he uses to describe particles flying through empty space—do not disintegrate over time. He believes this stability justifies replacing particles with wave packets in the theory.

However, that stability is the exception, not the rule. Max Born, being better at calculations, applies Schrödinger's theory to a more complex scenario: two particles in collision. And the result is very different. According to Born's calculations, the two wave packets should dissipate on colliding — spreading out in all directions through space, like waves on a pond when a stone is thrown into the water. If Schrödinger's theory were right, the particles would dissolve after colliding. But that doesn't fit with what we see happen every day in the world around us. Particles collide without dissolving.

If particles really are concentrations of waves, they must ultimately act in correspondence with what physicists and other people can see and measure. That's precisely what Niels Bohr meant when he formulated his correspondence principle. The description of a collision in terms of quantum mechanics must merge into a description in terms of classical physics; it must fit with the everyday intuitions through which we interpret the world we live in. Individual particles can't spread out through space. They have a specific position. After two particles collide, they fly apart again — not just in any old direction, but in well-defined trajectories, as can be observed in the case of Compton scattering. This is the behaviour of particles, not waves, no matter how long we observe them for.

Born ponders the significance of the results he's calculated. And he comes up with an elegant explanation. The waves that propagate from the point of collision are not the actual particles, but their probabilities.

This means Born has now switched sides, from using Heisenberg's equations to using Schrödinger's. But he does not switch to supporting Schrödinger's interpretation of reality. He remains convinced that particles can't just be abolished. 'However, it is necessary,' he wrote in his paper entitled 'On the wave mechanics of collisions', 'to drop completely Schrödinger's physical pictures that aim at a revitalisation of the classical continuum theory, to retain only the formalism and to fill that with new physical content'.

How can the two sides be reconciled? Max Born comes up with a trick that allows him to retain both particle theory and Schrödinger's equation. It lies in reinterpreting the nature of the wave function. Born no longer interprets waves as the particles themselves, but as their probabilities. Where the wave is highest, so is the probability of locating the particle. Where it is lower, there is less chance of finding the particle.

If Born is right, Schrödinger's equation describes something completely new. For an electron, for example, it does not describe its mass or charge distribution, as Schrödinger imagined, but the probability of finding the electron in one location or another.

But this reinterpretation of the wave function comes at a price. Born has 'unrealised' waves. For Born, waves are now nothing more than probability distributions, mere mathematical constructs.

Being the dispassionate mathematician he is, Born has no qualms about jettisoning waves. His main concern is that the theory must fit with the experimental results. During his time in the US, he laboured to find a way to describe collisions between atoms using cumbersome matrix mechanics. Now back in Germany, he produces two groundbreaking articles on the topic within a very short space of time, with the aid of wave mechanics. The first is only four pages long. It arrives at the editorial office of the journal *Zeitschrift für Physik* on 25 June 1926, bearing the title 'On the quantum mechanics of collision processes'. In it, Born writes, 'I myself am inclined to give up determinism in the world of atoms.' Ten days later, he posts his second, more thorough and more detailed article to the *Zeitschrift für Physik*.

Erwin Schrödinger is appalled. He denies the existence of particles altogether, and now here is Born using Schrödinger's theory to save particle theory. And, in doing so, rocking the eternal principle of physics—determinism. Schrödinger wants his equation to apply to things that are material, things that can be observed—such as the flowers, chairs, books, and people that surround us. Not something abstract like probability distributions.

The universe as Isaac Newton saw it is a paradise of pure determinism. There is no element of chance to it, only ignorance of cause and effect. Every moving particle has a specific location in space at a specific point in time, and a specific velocity. The forces at work on the particle determine how its location and speed change over time. This is also true for gases and liquids that contain a very large number of particles—at least, it's true in principle. In practice, it is impossible to observe every particle in such a huge mass of them. So the only option that was available to physicists such as James Clerk Maxwell and Ludwig Boltzmann was to explain the properties of gases using probabilities and statistics. For them, the need to use probability was a regrettable consequence of humans' ignorance in a deterministic universe that developed in strict adherence to the laws of nature. They thought that, if the human intellect were great enough to enable us to know the state of the universe, along with all the forces acting on it, at any given point in time, we would be able to calculate any possible future state of the universe. In classical physics, determinism and causality—the principle that every effect must have a cause—are connected as if by an umbilical cord. Hence, causality gives birth to determinism.

'Here the whole problem of determinism comes up,' Born writes in the paper that will go down in history as the founding document of the probability interpretation of quantum mechanics. 'From the standpoint of our quantum mechanics, there is no quantity which in any individual case causally fixes the consequence of the collision; but also experimentally we have so far no reason to believe that there are some inner properties of the atom which condition the definite outcome for the collision. Ought we to hope later to discover such properties (like phases or the internal atomic motions) and determine them in individual cases? Or ought we to believe that the agreement of theory and experiment—as to the impossibility of prescribing conditions for a causal evolution—is a pre-established harmony founded on the non-existence of such conditions? I myself tend to abandon determinism in the atomic world.'

This is unprecedented. Without batting an eyelid, Max Born has challenged a 300-year-old principle of science. Determinism is the linchpin of classical physics. What meaning can the principle of cause and effect now have when there is no longer a direct route from one to the other? How can meaningful physics even be possible, when everything is just a mere probability?

When two billiard balls collide, they can rebound in almost any direction. And when an electron collides with an atom, it can also rebound in almost any direction. But there is a difference between these two cases, Born argues. The movements of the billiard balls are already determined before the collision takes place. They are calculable. However, that's not the case for collisions at the atomic scale. Physics theory cannot answer the question of how particles move following a collision. It can only answer the question of how probable a given motion state is following the collision. It is impossible to predict where the electron will fly off to. All physicists can do is to calculate the probability that its direction of movement will be within a given angular range. This is what Born means when he writes about filling Schrödinger's equation with 'new physical content'.

The 'new physical content' means that the wave function has no physical reality; it exists only in the intermediate world of the possible, and describes only abstract possibilities such as the possible angle of the direction in which an electron will be scattered after a collision. The values of the wave function are complex numbers; their squares are real numbers. Born's claim is that those real numbers inhabit the realm of the possible.

Born's reinterpretation of the wave function paves the way for what will become known as the 'Copenhagen interpretation' of quantum mechanics. Soon after, Niels Bohr postulates that tiny objects such as electrons do not exist in any specific location when they are not being observed or measured. Between two measurements, they exist only in that ghostly intermediate world of the possible, which the wave function describes. It is only when the electrons are measured or observed that the 'wave function collapses', one of the 'possible' states of the electron becomes its 'real' state, and the probability of all other states becomes zero.

Thus Born reinterprets Schrödinger's wave function as a probability wave: a real wave has become an abstract one. He abandons the old determinism, but with a reservation: 'The motion of particles follows probability laws, but the probability itself propagates according to the law of causality.'

In the time between the first and second papers, Born realises the extent of the revolution he is sparking. He's not only 'probabilised' physics, but he's reinterpreted probability itself. The type of probability used by Maxwell and Boltzmann in their calculations is the probability of the unknown. To say after an exam, 'There's a 50-50 chance I passed' does not mean that your grade still depends on chance. Your grade has already been decided, but you don't yet know what it is. 'Quantum probability', as Born conceives it, can't simply be eliminated by additional knowledge. It's part of the fabric of reality at the atomic scale. The fact that it's impossible to predict whether a certain atom in a radioactive sample will decay, although it is certain that one of the atoms will decay, is not due to a lack of knowledge; it is a result of the nature of the quantum laws that govern radioactivity.

Before, reality came first and physicists then calculated probabilities. Now, probability comes first and reality then emerges. It appears to be a little trick, but it is a giant leap for physics.

And almost no one seems to care. Born's interpretation creeps into quantum mechanics without causing much of a stir. 'We were so accustomed to statistical considerations that it didn't seem particularly important to us when we went one level deeper,' Born later remarks. Of course, other quantum physicists say to themselves, Schrödinger was wrong about the nature of his waves. It's all about probabilities; we've always known that. Heisenberg claims to have known all along that he was calculating with probabilities, although he never wrote it in any of his publications. Textbooks on quantum mechanics often explain the probability interpretation without attributing it to Born, which is to become a source of annoyance to Born for the rest of his life.

Erwin Schrödinger and Albert Einstein are also annoyed. Schrödinger is vehemently opposed to what Born has done with his equation. He denies 'that such an individual event is absolutely random, i.e. completely indeterminate', in a letter to Wilhelm Wien dated 25 August 1926. Schrödinger also rejects '*a limine*', that is, from the outset and without further examination, Bohr's opinion that a space-time description of atomic processes is impossible. His reason is that 'physics does not consist only of atomic research, science does not consist only of physics, and life does not consist only of science. The aim of atomic research is to fit our empirical knowledge concerning it into our other thinking. All of this other thinking, insofar as it concerns the outer world, is active in space and time.'

Schrödinger is sure that if Born were right, quantum leaps would inevitably make a comeback and causality would be under threat. 'I don't need to eulogise over the beauty and clarity of your mathematical developments of the interference problem,' he writes to Born in November 1926, 'but I have the impression that you and others who generally agree with your views are too deeply in thrall to those concepts (such as stationary states, quantum leaps, etc.) which have gained civil rights within our thinking over the past twelve years, to do full justice to any attempt to escape that pattern of thinking.'

This is the point at which Schrödinger becomes a dissident. The theory for which he provided the most important equation is no longer his theory. Schrödinger adheres to his interpretation of wave mechanics, and to his efforts to explain atomic phenomena plainly, for the rest of his life. 'I can't imagine that an electron hops about like a flea,' he is known to say.

MUNICH, 1926

A turf war

Germany, 1926. The country is on the rise again. Hyperinflation has been curbed. More goods are passing through the ports. New shipping canals, power plants, and harbour facilities are being built. Not all the coal produced by Germany's mines is now being sent abroad as part of the war reparations, and the trains are moving again without fear of grinding to a halt due to lack of fuel. Wealthy Germans are buying furs in Paris, drinking champagne in Arosa, and motoring through the Upper Bavarian hills in their automobiles. Berlin is also coming back to life. The grey, dusty capital of a defeated country is becoming a lively, wild cosmopolitan city. Tourists have started to arrive again. Americans are curious to see what has become of the land of their forefathers. English visitors seek pleasure, free of the shackles of their class-ridden society. While the days are dominated by old Prussian prudery, the nights see Josephine Baker dancing in the bars along the Kurfürstendamm boulevard. Berlin's nightclubs cater to every sexual orientation—even while the Prussian law punishing sexual acts between men remains in force. Double standards are rife in 1926 Germany when it comes to morality.

Politically, the country is in a state of precarious equilibrium. Hitler's beer hall putsch has failed. Communists are mutinying in Saxony, while separatist sentiments are gaining ground in the Rhineland. The government cannot rely on the loyalty of the armed forces. Some people fear the country will break up; others hope it will.

Germany largely has one man to thank for its recovery: Gustav Stresemann, the fat-necked foreign minister with his casual-cut suits. He introduced the Rentenmark, a new currency to counter hyperinflation, which was secured by government and commercial bonds. He also persuaded his fellow government ministers to accept the Dawes Plan — named after the American financial expert Charles Dawes — that allowed Germany to pay its war reparations without ruining its economy.

Not even ten years after the end of the First World War, Germany is well on the way to becoming the second-most powerful industrialised nation in the world. The Locarno Treaties, signed in London on 1 December 1925, pave the way for Germany's political rehabilitation, allowing the Empire once again to manufacture aeroplanes and airships, which had been forbidden by the Treaty of Versailles. This rings in the age of the zeppelin. In 1926, the League of Nations admits Germany as a member. German scientists are no longer pariahs; their fellow scientists around the world are back on speaking terms with them. German regains its status as the international language of science.

On the afternoon of 23 July 1926, a sunny day in a mostly rainy summer, the future of physics is under negotiation. The Austrian physicist Erwin Schrödinger, having travelled from Zurich via Stuttgart, Berlin, Jena, and Bamberg, is in the auditorium of Munich's Ludwig-Maximilian University, delivering a lecture on 'The fundamental concepts of wave mechanics'.

Until just a few months ago, Schrödinger was a scientific outsider. Zurich is far removed from the great stages of quantum physics — Berlin, Göttingen, and Munich. But when wave mechanics starts to spread like wildfire among physicists in the spring and summer of 1926, Schrödinger becomes a sought-after speaker, and everyone wants to discuss his theory with him face to face.

Schrödinger is quick to accept an invitation from Arnold Sommerfeld and Wilhelm Wien to deliver two lectures in Munich. The first takes place on 21 July 1926 as part of Sommerfeld's 'Wednesday colloquium'. Schrödinger reels off his talk effortlessly, and receives friendly applause. His second 'Fundamental concepts of wave mechanics' lecture, given two days later at the regional association of the Physical Society, goes off less smoothly, and the reason for that is Werner Heisenberg.

The lecture theatre is buzzing with the murmur of voices. Men, and a few women, squeeze between the rows of seats, scan the room, exchange greetings, nod, wave, and shake hands with each other. Sunlight streams through the high windows of the hall, flooding the room.

The most eminent figures of the scientific world take their seats in the front row, with their starched collars, long frockcoats, and pomaded hair. They include Wilhelm Wien, director of the Institute of Physics and Metronomics and current rector of the university, and Arnold Sommerfeld, head of the Institute for Theoretical Physics. Students are relegated to the back seats. Among them is a young American chemist by the name of Linus Pauling, who had arrived in Europe the previous April on a Guggenheim fellowship. Pauling is ten months older than Heisenberg, but is far less advanced along the path towards a glorious scientific career.

Werner Heisenberg, with his shock of wiry blond hair, his cheeky, boyish face, and his bright, clear eyes, is late to enter the lecture hall. At the age of twenty-four, he already rules the roost of quantum mechanics. He came up with his theory, which he simply and unapologetically calls *the* theory of quantum mechanics, a few months before Schrödinger developed his. He, not Schrödinger, should really be the one standing at the front, delivering a lecture. Heisenberg has cut short his hiking trip to Norway and hurried across Europe to defend his territory in Munich. He had travelled north to escape his hay fever, to climb mountains, and to 'steamroll', as he puts it, other quantum physicists with his theory. He's spent the previous weeks camping by the side of Lake Mjøsa, thinking about quantum mechanics during the white Norwegian nights, and applying the theory to the long-standing puzzle of the spectral lines of helium, while hiking through the valley of Gudbrandsdal and along the Sognefjord. He's now returned to Munich full of self-confidence. His face is tanned from the long, sunny days of the Scandinavian summer.

Less visible than Heisenberg's tan are the conflicts between many of those present. Wien and Sommerfeld are keen to return the old order to physics, to bring it back to a state that would please Isaac Newton and James Clerk Maxwell. To effect a return to the kind of physics that renders the world not only predictable, but also understandable. Physics that not only produces equations, but also describes the world as it really is.

That's the hope that the old masters of physics place in Erwin Schrödinger. It is this hope that motivated Wien and Sommerfeld to invite him to come to Munich. Their kind of physics is under threat, and that threat is called Werner Heisenberg. His matrix mechanics is a formalistic monster, an affront to physical intuition. Not even position and motion retain their meaning in matrix mechanics. Heisenberg has derived his matrix mechanics in a long, lonely struggle by banishing any idea of 'visualisability' from his thoughts. He has developed a kind of formalism that raises many philosophical questions, and which is so strange and cumbersome that most physicists would have to master a whole new branch of mathematics to be able to understand it. It leaves them looking like schoolboys, and only a handful of physicists manage to wrap their minds around Heisenberg's edifice of equations.

A few months after Heisenberg formulated his matrix mechanics, Schrödinger developed his wave mechanics during his winter vacation in Arosa. He discovered a simple way of describing the processes that take place inside the atom. It has the same explanatory power as Heisenberg's theory, and promises to make those processes as easy to picture and understand as the ripples on a pond, using the kind of equations that physicists have known and used for centuries.

It's as if Schrödinger has turned on a light to illuminate the darkness of the quantum world; as if physicists can now see what they are dealing with. The promise of Schrödinger's wave mechanics is to show that the smallest building blocks of nature are not the mysteries they are thought to be by Bohr, Heisenberg, and their ilk. Rather, they obey the old, familiar, understandable laws, as they are taught to beginner students of physics.

The conflict between Schrödinger and Heisenberg is not mathematical in nature. Formally, their theories are of equal worth;

one can be converted into the other, as shown by Erwin Schrödinger himself, and by Wolfgang Pauli working in Hamburg, and by Carl Eckart in California. This is not about equations. It's about interpretations. Schrödinger understands the atom to be a vibrating system. Yes, he understands it. He paints a visual picture of the physical processes—which is exactly what Heisenberg deems impossible. Heisenberg doesn't 'understand'; he calculates. It is precisely the rejection of understanding that formed the key step towards the development of his matrix mechanics. And now Schrödinger is his competition, with his unashamedly understandable wave mechanics.

It's a competition for the 'soul' of physics, as Schrödinger puts it. The goal of physics is to make the world more familiar to us, not less, as Heisenberg's matrices do. Schrödinger is pleased by 'the human proximity of all these things', thanks to his gently flowing waves. He calls his own theory 'physical', thereby tacitly implying that Heisenberg's is 'unphysical'. The jumping from one state to the next that Heisenberg's matrices describe is anathema to him.

Just how much Schrödinger despises it can be seen in his paper 'On the relation between the quantum mechanics of Heisenberg, Born, and Jordan, and my own', published in *Annals of Physics* in May 1926. In that article, he is outrageously frank in describing himself as being 'discouraged, if not repelled', by the 'algebra and by the want of perspicuity' in Heisenberg's theory.

The feeling is entirely mutual. 'What Schrödinger writes about the visualisability of his theory ... I consider to be crap,' writes Heisenberg in a letter to Wolfgang Pauli. And Pauli refers to Schrödinger's interpretation as a 'local Zurich superstition'—a phrase that soon catches on. Schrödinger is offended when he hears it. Pauli apologises to 'my dear Schrödinger', and asks him 'not to see it as a personal unkindness to you, but as an expression of the objective belief that quantum phenomena display such aspects in nature that cannot be described by the concepts of continuum physics (field physics) alone ... My dear Schrödinger, you have a pretty theory but, and don't take this amiss, it doesn't fit with the world we live in.'

It's true: none of it is meant personally, but what good is that when dealing with people whose personalities are so thoroughly defined by the fact that they are scientists? Heisenberg and Schrödinger have nothing against each other personally. But as scientists, they have a lot against each other. Each is very familiar with the other's theory. Now, in this Munich lecture theatre on a July afternoon, they meet personally, coming face to face for the first time.

The Swiss astrophysicist Robert Emden acts as the host, greeting the audience and thanking Schrödinger for coming. Schrödinger steps up to the lectern, cutting a small, slender figure in his linen suit and bowtie. His skin has the bronzed tone of one who spends a lot of time in the mountains. Mountain air helps his tuberculosis—a disease he has battled all his life, and which will eventually kill him.

Schrödinger begins his talk, his voice melodic and Viennese-accented. He presents the equation that has immortalised his name. It's one of the most beautiful, but also strangest, equations ever devised by the human mind. It describes the movements of electrons in the energy field of an atomic nucleus so elegantly that some of Schrödinger's fellow physicists describe it as 'transcendental'.

It seems like magic. Schrödinger has preserved the formal power of quantum mechanics while simultaneously describing a world that the old-school physicists crave—a world that glides smoothly, steadily, and understandably from one state to another, free of jumps and discontinuities.

He is convinced that his wave equation is not just an abstract construct. It describes an actual, real-world wave. But what is the wave's medium? What is this wave's 'water'? On this point, Schrödinger wavers. It may be something like the distribution of an electron's charge over space. An electron, he believes, is not actually a particle at all, but a particle-like wave.

Erwin Schrödinger is an impressive speaker, and he soon has the audience in the hall on his side. Wilhelm Wien, the experimental physicist and rector of the university, had been the first person Schrödinger revealed his discovery to, written on a postcard sent to him from Arosa in December 1925. He hears what he had been hoping to hear: that Schrödinger has deciphered the atom, as Wien finds, 'in a pleasant way, which is connected most closely to the classical theory'. Wien is relieved, as 'the previous state of the theory' had 'become unbearable' for him. Reassured, he states that 'quantum theory has now been merged with classical theory after all', and that young physicists 'can no longer splash around in the mire of whole and half quantum discontinuities'. He dismisses Bohr's atomic model, in which electrons jump from orbit to orbit, as 'atomic mysticism', and denounces Heisenberg's matrix mechanics as a formal monster, far removed from the experimental reality that was Wien's scientific home. Both were aberrations of scientific research, according to Wien, which he hopes will now be banished from physics by Schrödinger's vibrating waves.

The American student Linus Pauling is so impressed by Erwin Schrödinger's ability to bring clarity to the quantum world that he decides to set all his previous research interests aside and try to emulate Schrödinger. For nights on end, he immerses himself in wave mechanics. A months later, he will even transfer to Zurich to study under Schrödinger.

However, Schrödinger is not able to convince everyone in the audience. 'I don't believe Schrödinger,' Sommerfeld admits to Pauli in a letter written a few days later. He praises Schrödinger's wave mechanics as a form of 'admirable micromechanics' that, however, 'doesn't even remotely solve the fundamental quantum conundrum'. In the lecture theatre, Sommerfeld keeps his misgivings to himself. He doesn't want to snub the guest he invited.

Heisenberg is also able to hold back—until the final moment, at the end of Schrödinger's second lecture. Then, during the follow-up discussion, he can't contain himself any longer. Quantum mechanics is his invention, and now here's Schrödinger presenting himself as the man who will save the world from it? He can't let that go unchallenged. This is his city, it's his theory, and this is his turf. This is where he went to school, studied, and earned his doctorate.

Heisenberg rises to speak. All eyes turn to him. He speaks out vehemently against Schrödinger. How can he claim atoms vibrate gently away to themselves, when so many experimental results have shown that they do not do so? Results demonstrate that abrupt jolts take place within them. And what about the photoelectric effect, the Franck-Hertz experiment, Compton scattering? Schrödinger's theory is unable to explain any of those phenomena. Not even Planck's law on radiation will fit into the theory, cries Heisenberg. And of course it never will, as none of those things can be explained without the existence of particles, discontinuities, and quantum leaps—all things that Schrödinger wants to get rid of.

Angered, Wilhelm Wien rises and reprimands Heisenberg. He can understand Heisenberg is upset that Schrödinger has dispatched with matrix mechanics and put an end to quantum leaps and all that nonsense. Wien adds that he's sure 'Professor Schrödinger' will answer all the outstanding questions soon. Neither of the two has forgotten that Wien wanted to fail Heisenberg in the oral examination for his doctorate three years earlier. It was only because Heisenberg's mentor, Sommerfeld, put in a good word for him that he was finally awarded his doctor's title. And now Wien sees an opportunity to fail Heisenberg after all. 'Young man, you must learn physics first,' he says, cutting Heisenberg off and gesturing for him to sit back down. 'He almost threw me out of the lecture theatre,' Heisenberg would later recall.

Schrödinger's response is more cautious, when Wien eventually lets him answer. He admits that Heisenberg's objections to wave mechanics are justified, but he's confident he will be able to clear them up very soon. Heisenberg is not prepared to let it drop, pointing out an even more fundamental problem. How can these waves be 'real' if they can't be observed? Why should we expect our common sense, trained, as it is, to deal with chairs, flowers, and our fellow human beings, to help us understand the subatomic world? That world can't be compared to anything we're familiar with—Heisenberg is sure of this, and his words are full of that certitude. The dark, deep core of the physical world can't be explained by a simple wave equation. No way. If it could, all his struggles with these mysteries would have been in vain.

This is not the easy home game that Heisenberg had hoped for. Sommerfeld remains politely silent. As for the others in the audience, the more heated Heisenberg's objections to Schrödinger get, the more sceptical they become. Why should we have to abandon common sense, that human instinct that has been proven and honed for centuries, in order to understand matter at the smallest scale? It's probably just envy on Heisenberg's part—and no wonder, as this competitor is threatening to steal his place in history.

A stricken Heisenberg slopes back to the house of his parents in Hohenzollernstrasse. He feels as if he's been sent off the battlefield before the fight has even begun. And he is worried 'to see that many physicists saw this interpretation of Schrödinger's as a deliverance'. Even Sommerfeld, Heisenberg laments, 'can't escape the persuasive power of Schrödinger's mathematics'. Nothing against Schrödinger, but that's not the way to get the better of quanta. 'Nice as Schrödinger is as a person, I find his physics to be odd,' he writes to Wolfgang Pauli. 'When you hear it, you feel twenty years younger.'

Heisenberg needs a few days to regain his composure. He gives vent to his frustration by writing letters. 'A couple of days ago, I attended two talks by Schrödinger, and since then I have been absolutely convinced that Schrödinger's physical interpretation of QM is wrong,' he writes to his Göttingen colleague Pascual Jordan. But he knows such conviction alone will not be enough in the face of the elegance and charisma of both Schrödinger and his mathematics.

Niels Bohr is also concerned when he hears of his assistant's public spat with Erwin Schrödinger. Resolving to bring the two rivals together, he invites Schrödinger to Copenhagen, 'so that we can discuss more deeply the still-unresolved issues of atomic theory within the closer circle of those who work here at the institute, which currently includes Heisenberg, as you know'. Schrödinger gratefully accepts the invitation. He's looking forward to 'being able to discuss the difficult, burning questions' with Bohr. But first he travels to South Tyrol to enjoy the fresh mountain air with his wife, Anny. Better to be in the best of health when he visits Bohr. The events in Munich will turn out to be only the opening scene in the drama to come.

COPENHAGEN, 1926

Exquisitely carved marble statues falling out of the sky

Cambridge, September 1926. Paul Dirac is setting off on a journey. Quantum mechanics is the hot new research topic that's the talk of the continent, and Dirac wants to be part of that conversation. Keen to find a way to explain how atoms work, he heads off on a tour taking in the luminaries of quantum physics in order to learn from them — Niels Bohr in Copenhagen, Max Born in Göttingen, and Paul Ehrenfest in Leiden.

The journey across the North Sea takes sixteen hours. The autumn storm season has begun. Dirac gets seasick, and spends most of the crossing throwing up. Never again? That's not the way Dirac thinks. He decides to spend more time on the stormy seas to cure himself of the weakness of seasickness. 'Dirac is rather like one's idea of Gandhi,' a colleague once said of him. 'He is quite indifferent to cold, discomfort, food, etc.' Dirac doesn't smoke or touch alcohol. His favourite drink is water. For supper, a bowl of porridge is enough for him. 'Of all physicists, Dirac has the purest soul,' says Niels Bohr.

That purity is part of Dirac's strength as a scientist, but it's a weakness when it comes to social interactions. Once, when a group of physicists are having tea together, Wolfgang Pauli adds so many lumps of sugar in his tea that his colleagues start to make fun of him. All except Dirac, who remains silent and serious. The other physicists jokingly ask Dirac what he thinks of 'the sugar lump problem'. After mulling it over for a while, Dirac answers, 'I think one lump of sugar is sufficient for Professor Pauli.' The conversation moves on, but after a couple of minutes, Dirac pipes up, 'I think one is enough for anybody.' Again, the conversation moves on. But, still stuck on the same train of thought, Dirac eventually says, 'I think the lumps are made in such a way that one is enough for anybody.'

Dirac is to spend four months with Bohr. There would be a lot for them to talk about, but Dirac is a taciturn character. 'This Dirac seems to know a lot of physics,' Bohr observes, 'but he never says anything.' Dirac spends a lot of time in Copenhagen thinking, which he prefers to do alone. Bohr reports that Dirac is 'the strangest man' ever to have visited his institute in Copenhagen.

For his part, Dirac is also fascinated by Bohr. He admires him, and is amazed by his sociable nature and his bumbling lecture style. 'People were pretty well spellbound by what Bohr said,' Dirac reports, while complaining that 'his arguments were mainly of a qualitative nature, and I was not able to really pinpoint the facts behind them. What I wanted was statements that could be expressed in terms of equations, and Bohr's work very seldom provided such statements.' Dirac conjectures that Bohr would make 'a good poet, because in poetry it is useful to use the words imprecisely'.

Dirac marvels at Bohr and admires him, but he doesn't worship him as a cult idol in the way the Dane has become accustomed to. And it is precisely that which earns the tall loner from England Bohr's respect. Bohr struggles to put his philosophical inspirations into words. Dirac seeks clarity and logical precision. He never speaks until he is completely sure of every detail of what he wants to say. His strength is his ability to capture the essence of nature with elegant mathematics. Bohr needs people like Dirac. And Dirac needs someone like Bohr. He knows that equations alone are not enough. They also require interpretation, and that's a job Dirac prefers to leave to others—like Bohr, Heisenberg, or Schrödinger.

Dirac conducts a large proportion of his conversations in Copenhagen with a repertoire of three utterances—'Yes', 'No', and 'I don't know.' He leads an extremely regimented life, almost like Immanuel Kant, working on his theories five days a week, leaving

Saturdays for his technical projects. On Sundays he goes hiking. Week after week follows the same pattern.

Working in his little study, Paul Dirac changes the world of quantum mechanics with two groundbreaking papers, both of which he authors in the space of twelve weeks.

The first paper is the one he will later call his 'favourite'. Since Schrödinger published his wave mechanics, there has been broad agreement that there is no room in physics for two quantum theories. There can be only one. In his favourite paper, Dirac proves that Heisenberg's and Schrödinger's formulations of quantum mechanics, which appear so fundamentally different at first glance, are equally valid and can be mutually transformed into each other. Others have shown this before—Schrödinger himself, Pauli, Eckart. But Dirac takes it a step further, discerning a mathematical structure that underlies both theories, which has never been discovered before. He calls this structure 'transformation theory'. It is even further removed from physical visualisability than Heisenberg's matrices. Dirac has no problem with mathematical abstraction. His theory provides a mathematical justification of Schrödinger's equation, which Schrödinger devised in an act of almost artistic creativity. Dirac's theory transforms poetry into truth.

A couple of weeks later, he publishes the second paper, which opens up a new area that will still be part of our fundamental understanding of physics almost a century later: quantum field theory.

After his stay in Copenhagen is over, Dirac travels on to Göttingen to visit Max Born, Werner Heisenberg, and Pascual Jordan. He lodges with a family in their Göttingen villa, which he shares with Robert Oppenheimer, Max Born's highly gifted, fast-talking, self-promoting doctoral student from America. Despite their great differences, Dirac and Oppenheimer become friends. While Dirac gets up early every day, Oppenheimer often doesn't go to bed until the early morning. Dirac loves equations; Oppenheimer loves poetry, and even writes poems of his own. 'I don't see how you can work on physics and write poetry at the same time,' Dirac tells Oppenheimer. 'In science, you want to say something nobody knew before, in words everyone can understand. In poetry, you are bound to say something that everybody knows already in words that nobody can understand.' What the two men share is their love of physics. 'Perhaps the most exciting time in my life was when Dirac arrived and gave me the proofs of his paper on the quantum theory of radiation,' Oppenheimer will later say. Until then, barely any physicists had understood Dirac's quantum field theory. To Oppenheimer, it is 'extraordinarily beautiful'.

The German physicists in Göttingen have more difficulty getting along with Dirac, whom they see as an oddball. All the Göttingen set are very able mathematicians, but Dirac's use of such esoteric maths is unusual even by their standards, especially combined with his engineering knowledge, which some of the Göttingen theoreticians feel is beneath them. That is not the German way of doing physics.

The few scientists who do understand Dirac's work have deep respect for him. No less a person than Werner Heisenberg, the now twenty-six-year-old creator of quantum mechanics, remarks only half-jokingly, 'I believe I shall have to give up physics. There is this young Englishman by the name of Dirac, who is so clever it is not worthwhile trying to compete in one's work with him.'

This is only the start of Dirac's great creative period. Over the coming eight years, he will make discovery after discovery—each more elegant than the last. His fellow mathematical physicist Freeman Dyson compares them to 'exquisitely carved marble statues falling out of the sky, one after another. He seemed to be able to conjure laws of nature from pure thought—it was this purity that made him unique.' Compared to Dirac, other quantum physicists look like amateurs. They get their theories wrong, make mistakes, and have to go back and correct themselves. Dirac's work is successful straightaway.

In 1928, Dirac develops an equation of immaculate beauty that bears his name—the Dirac equation. It looks like this: $(i\partial\!\!\!/ - m)\psi = 0$. Succinct, perfect, a fitting monument to its taciturn creator, and perhaps the most beautiful equation in physics.

At the time Dirac puts his equation down on paper, physics rests on two pillars: Albert Einstein's special theory of relativity, and Erwin Schrödinger's quantum mechanics. But physicists have found no way to unite these two world-changing discoveries. Even Schrödinger has failed to find a way. Paul Dirac finds it. With one equation, he succeeds in reconciling the theories of Einstein and Schrödinger.

If ever there were a 'world equation', this is it. With this equation, Dirac describes the electron—the particle whose behaviour defines the whole of chemistry and that determines how humans see the world. When light hits the human eye, it excites the electrons in the optic nerves. The Dirac equation governs that process. Furthermore, the equation also applies to the building blocks of matter that have been discovered since—even down to the level of quarks and muons.

Dirac's equation also explains electron spin, which Wolfgang Pauli had postulated a year earlier without being able to offer a theoretical explanation for it. To provide this explanation, Dirac makes use of an abstract structure taken from a very esoteric branch of mathematics that was previously the preserve of a few cloistered algebraists. It will later come to be called a 'spinor'.

When Dirac applies his equation to the electron, it yields not one solution, but two—one with positive energy and the other with negative energy but positive charge. What could that be—negative energy? Dirac keeps recalculating, but he can't get rid of that negative solution. Has he happened upon a description of the proton? That doesn't fit. It's as if the electron had a previously unnoticed mirror image, as if there were 'holes' in the world that behave in a similar way to the known particles, but with the opposite sign. Dirac has stumbled upon antimatter, the stuff that made up half the universe shortly after the Big Bang. With his mathematical argument he has discovered an entire half-world.

At first, no one will believe him. In his seminar, Enrico Fermi 'condemns' Dirac to the bastinado—that is, a foot-whipping—for the absurdity of the idea of antimatter. But just a few years later, the positron is detected by experiment and has the very properties predicted by Dirac. Werner Heisenberg calls the discovery of antimatter 'probably the greatest of all leaps that physics has made in our century'.

But Paul Dirac doesn't stop there. Using a topological argument, he predicts the existence of magnetic monopoles. Until now, it has been a hard-and-fast rule of physics that magnetic poles can only exist in pairs. Where there's a north pole, there must also be a south pole. Just as a sheet of paper must have two sides—it can't have a front without a back. However, in topology you can have an abstract strip of paper that looks like a double-sided sheet from up close, but in fact only has one side—the Möbius strip. A magnetic monopole is like a magnetic Möbius strip.

There are brilliant mathematicians. There are excellent physicists. And then there's Paul Dirac, who is both at the same time—an unworldly mathematician and a worldly engineer who boldly makes the connection between the square root of a vector and the existence of an entire mirror world.

In 1932, Paul Dirac takes up the renowned Lucasian professorship of mathematics at the University of Cambridge—a chair once occupied by Isaac Newton himself.

COPENHAGEN, 1926

A game with sharpened knives

On 1 October 1926, Erwin Schrödinger arrives in Copenhagen by train, his confidence boosted by his visits with Arnold Sommerfeld in Munich and Albert Einstein in Berlin. Niels Bohr, already considered the grand old man of quantum physics at the age of forty-one, is waiting for Schrödinger on the station platform. This is the first time the two men have met. Perhaps it's too late. Only a few months earlier, they would have been allies in the battle for a wave theory of atomic phenomena. But not anymore. Bohr has lost the faith. Schrödinger remains true to the wave theory. Ever since Max Born reinterpreted his wave equation, Schrödinger has gone from being a defender of quantum mechanics to one of its critics. Bohr would like to hear that criticism from Schrödinger personally.

Bohr keeps the exchange of pleasantries short, and even before they have left the station he launches into a relentless interrogation of Schrödinger. It will be eight days before Bohr gives his guest any respite. On 4 October, Schrödinger delivers a lecture on his wave mechanics at the 'Physical Association' of Bohr's institute. After that, Bohr resumes his onslaught.

Margrethe and Niels Bohr are kind and attentive hosts. They put Schrödinger up in their guest room so that Erwin and Niels can spend as much time as possible together. However, that time is not spent peacefully, but rather in a kind of scientific duel. Schrödinger defends his 'standpoint of visualisable pictures', while Bohr disputes that very visualisability. Again and again, he tries to shake the foundations of Schrödinger's standpoint, to pick fault with his reasoning. Schrödinger dodges, counters. Neither will back down. 'Science is a game,' says Schrödinger, 'but it's a game with reality, a game with sharpened knives.'

Werner Heisenberg, who has just started his job as Bohr's assistant, is also present. Heisenberg is four years younger than Schrödinger, but he is always one step ahead of him strategically. That was the case with their theories, and it was the case with Planck and Einstein in Berlin, and now it's the case with Bohr in Copenhagen. Schrödinger is the hare; Heisenberg, the tortoise.

Heisenberg does not get involved in the dispute between the two men. It's only been two months since he clashed with Schrödinger in Munich, and that didn't end well for him, so he simply watches and listens. He hears Schrödinger speculate that it might still be possible to derive Planck's law on radiation, revealed back in 1900, without resorting to the use of quanta. Impossible, responds Bohr. Quanta are here to stay.

Heisenberg barely recognises his boss. Normally so kind and considerate, Bohr now seems to Heisenberg almost 'like a relentless fanatic, who was not prepared to concede a single point to his interlocutor or to allow him the slightest lack of precision'.

Two of the world's leading physicists confront each other, but neither gives any ground, not even a millimetre. Behind the words they relentlessly fling at each other, Heisenberg can hear each scientist's deep conviction.

According to Heisenberg's memoirs, Schrödinger firmly believes that the movement of electrons must be governed by a set of laws. Anything else would not be physics; it would be a capitulation by physical science. 'You surely must understand, Bohr, that the whole idea of quantum jumps necessarily leads to nonsense. It is claimed that the electron in a stationary state of an atom first revolves periodically in some sort of an orbit without radiating. There's no explanation given of why it should not radiate; according to Maxwell's theory, it must radiate. Then the electron jumps from one orbit to another one and thereby radiates. Does this transition occur gradually or suddenly? If it occurs gradually, then the electron must gradually change its rotation frequency and its energy. It's not comprehensible how this can give sharp frequencies for spectral lines. If the transition occurs suddenly, in a jump, so to speak, then indeed one can get from Einstein's formulation of light quanta the correct vibration frequency of the light, but then one must ask how the electron moves in the jump. Why doesn't it emit a continuous spectrum, as electromagnetic theory would require? And what laws determine its motion in the jump? Well, the whole idea of quantum jumps must simply be nonsense.'

Bohr feigns a concession, but, in fact, gives no ground at all on this point. 'Yes, in what you say, you are completely right. But that doesn't prove that there are no quantum jumps. It only proves that we can't visualise them. That means that the pictorial concepts we use to describe the events of everyday life and the experiments of the old physics do not suffice to also represent the process of a quantum jump. This is not so surprising when one considers that the processes with which we are concerned here cannot be the subject of direct experience, and our concepts do not apply to them.'

Schrödinger can tell that Bohr is trying to lure him into the deep waters of philosophy, and begins to backpedal. 'I don't want to get into a philosophical discussion with you about the formation of concepts—that's a matter for the philosophers to deal with later—but I should simply like to know what happens in an atom. It's all the same to me in what language you talk about it. If there are electrons in atoms, which are particles, as we have so far supposed, they must also move about in some way. At the moment, it's not important to me to describe this motion exactly; but it must at least be possible to bring out how they behave in a stationary state or in a transition from one state to another. But one sees from the mathematical formalism of wave or quantum mechanics that it gives no rational answer to these questions. As soon, however, as we are ready to change the picture, so as to say that there are no electrons as particles but rather electron waves or matter waves, everything looks different. We no longer wonder about the sharp vibration frequencies. The radiation of light becomes as easy to understand as the emission of radio waves by an antenna, and the formerly unsolvable contradictions disappear.'

Bohr disagrees calmly, but firmly, 'No, unfortunately that's not true. The contradictions do not disappear; they are simply shifted to another place. You speak, for example, of the emission of radiation by the atom, or, more generally, of the interaction of the atom with the surrounding radiation field, and claim that by assuming there are matter waves but no quantum jumps, the difficulties will disappear. But just consider the thermodynamic equilibrium between the atom and the radiation field, and of Einstein's derivation of Planck's radiation law. For the derivation of this law, it is essential that the energy of the atom have discrete values and change discontinuously; discrete values of the natural frequencies do not help at all. You can't seriously wish to question the entire foundations of quantum theory.'

However, that's precisely what Schrödinger does wish to do. 'Naturally, I do not maintain that all these relations are already completely understood. But you don't yet have any satisfactory physical interpretation of quantum mechanics either. I don't see why one shouldn't hope that the application of thermodynamics to the theory of matter waves will eventually lead to a good explanation of Planck's formula—which, however, will differ somewhat from the previous explanations.'

Bohr persists in contradicting him, 'No, one cannot hope for that. For we have known for twenty-five years what the Planck formula means. And besides, we see the discontinuities, the jumps, quite directly in atomic phenomena, perhaps on a scintillation screen or in a cloud chamber. We see a flash of light suddenly appear on the screen, or an electron suddenly move in the cloud chamber. You can't simply wave away these discontinuous phenomena as though they didn't exist.'

Schrödinger raises his hands in a gesture of impatience. 'If we are still going to have to put up with these damn quantum jumps, I am sorry that I ever had anything to do with quantum theory.'

Bohr now knows that he has won. He signals this to Schrödinger by adopting a more conciliatory tone. 'But the rest of us are very thankful that you have, and your wave mechanics, in its mathematical clarity and simplicity, is a gigantic advance over the previous form of quantum mechanics.'

And so the discussion continues, every day from morning till night. For Bohr, this is his normal way of doing science, but for Schrödinger it feels as if he's caught in an inescapable, Kafkaesque interrogation. His constitution being generally weak, Schrödinger's strength fails him after a few days. He's confined to his bed in the Bohrs' guest room with a feverish cold. Mrs Bohr takes care of him, bringing him tea and cake, while her husband sits on the edge of the bed and continues their argument. 'But surely, Schrödinger, you must see ...' Schrödinger stares at him, his eyes bright with fever. Neither scientist is able to convince the other. Each has little more to offer than his own conviction that he is right.

Their disagreement is not about facts that can be verified by experiment or by mathematical arguments. The facts are self-evident. This is a dispute over the interpretation of those facts. Both physicists have a deep-rooted conviction; neither will give up even the tiniest bit of ground without a fight. Neither leaves any weakness in his opponent's argument unexploited. But neither has a coherent interpretation of quantum mechanics.

There is no consensus in sight. Their perspectives are just too different. Schrödinger sees quantum mechanics as a seamless continuation of classical physics. Bohr believes breaking away from classical realities is precisely what is needed. There can be no return to the old concepts, to steady motion and continuous orbits. Quantum jumps are not going to go away, whether Schrödinger likes them or not. Schrödinger becomes increasingly angry and frustrated, unable to find an answer to Bohr's quiet, incessant attacks.

An exhausted Schrödinger boards the train back to Zurich. He's glad to finally escape the constant questioning. Bohr has ground him down, but not convinced him. No, Bohr has not refuted his 'standpoint of visualisable pictures', although, he admits to himself, he may now have to rethink a few details. 'It is absolutely probable,' he confesses, 'that here and there a wrong path was taken that must now be abandoned.'

Schrödinger describes his adventure in Copenhagen in a long letter to Wilhelm Wien. Initially, he focuses on the positive aspects. 'It was very nice that I was able to become thoroughly acquainted with Bohr, whom I had never met before, in his own surroundings, and to talk with him for hours about these matters which are so very dear to all of us.' He found his host 'profoundly likeable'. 'There will hardly come in the near future another man who scores such immense external and inner successes, to whom one pays homage, in his sphere of activity, nearly as to a demigod in the entire world, and who remains at the same time really bashful and shy like a candidate of theology.' However, Schrödinger then moves on to Bohr's 'current approach to the atom problems', which he considers 'really very remarkable. He is completely convinced that any understanding in the usual sense of the word is impossible. Therefore the conversation is almost immediately driven into philosophical questions, and soon you no longer know whether you really take the position he is attacking, or whether you really must attack the position that he is defending.' While admitting that his own view is not yet fully developed, he contests Bohr's view that 'visualisable wave pictures work as little as the visualisable point-particle models'. He sums up his own point of view thus: 'For me, the comprehensibility of the external processes in nature is an axiom. The facts of experience cannot contradict each other. If that seems to be the case, then some theoretical connections must be untenable. To seek these most untenable parts in the currently most tenable concepts seems to me to be very premature; I mean in the completely general conceptions of space and time and the connection of the interaction of neighbouring space-time points—all concepts that had been preserved, for example, even in general relativity theory.'

But, Schrödinger adds, 'It is not so easy to be sure how Bohr means things, partly because he often speaks for minutes on end in a dreamily visionary and really very unclear way, and partly because he is so considerate.' Bohr politely acknowledges the achievements of his verbal sparring partner, but Schrödinger doesn't 'care at all for this whole play of waves, if it should turn out to be nothing more than a comfortable computational device for evaluating matrix elements'. The two do not see their scientific differences as a reason to dislike each other personally. 'In spite of all these theoretical points of dispute,' Schrödinger writes at the close of his letter to Wien, 'the relationship with Bohr, and especially Heisenberg, both of whom behaved toward me in a touchingly kind, nice, caring and attentive manner, was totally, cloudlessly amiable and cordial.' Physical distance and a couple of weeks of convalescence temper Schrödinger's memory of his tortuous stay with the Bohrs.

Having a common opponent brings the Copenhagen set closer together. Heisenberg refers to himself, Bohr, and their likeminded fellow physicists as 'We Copenhageners'. After the war of words with Schrödinger, Niels Bohr and Werner Heisenberg are 'very certain' they are 'on the right track' with their campaign of opposition to the belief in the normal reality of quantum physics: they believe that the processes that take place in atoms cannot be described using traditional concepts of position and motion. But now they have had a foretaste of the kind of battles that lie ahead.

COPENHAGEN, 1927

The world goes fuzzy

Copenhagen, autumn 1926. The feeling of exhilaration that Heisenberg had after Schrödinger's visit has now faded. Now he's the one who constantly has to face Bohr's barrage of questions. He feels he has no place to retreat to. Only a staircase separates his office from Bohr's. Bohr feels free to appear at Heisenberg's door at any time, filling the office with pipe smoke and holding forth with one of his dreaded monologues.

Niels Bohr and Werner Heisenberg are striving to come up with a joint 'Copenhagen' interpretation of quantum mechanics. They strive with each other and against each other. The equations are there: fixed, mathematical, logical. But what use are they when the old concepts of position and velocity no longer apply? Bohr and Heisenberg soon realise that they do not agree as closely as they thought.

Heisenberg would prefer to avoid such interpretative questions, believing the theory speaks for itself. But he doesn't want to leave all the interpretation to Bohr. This is his theory, he invented it, and he has the right to have a say in its interpretation.

Bohr, on the other hand, considers Heisenberg to be an immature thinker. His mind is sharp and creative, but philosophically shallow. The theory has been formulated; now what's needed is wisdom. And he, Bohr, is the right man to provide it.

The two men spend many hours together every day, Bohr relentlessly lecturing, while an increasingly frustrated Heisenberg tries to get a word in. On many an evening, they continue their debates while taking a walk through the Fælledparken gardens next to the institute.

Heisenberg is unable to find peace, even in his well-appointed flat under the eaves of the Bohr Institute, with its view over the leafy park. Sometimes Niels Bohr will turn up at his door late in the evening, a bottle of sherry in hand, wanting to run through some thought experiment or other with him, to clear up an earlier misunderstanding or to restart an earlier discussion. Bohr has no concept of time off. If he has something to say, he must say it straightaway.

However, the two are of one mind when it comes to opposing Erwin Schrödinger. Schrödinger describes quantum systems as gently rolling waves. He considers particles and their jumps to be illusions. Bohr and Heisenberg believe he's on completely the wrong track, enticed by wishful thinking and nostalgia for a long-gone era of physics. As for Heisenberg, now that the old paths can no longer be trodden, he's in favour of a radical change of direction and a total reconsideration of everything. He believes physicists will need to learn a whole new language in order to understand quantum phenomena.

Bohr finds Heisenberg's aspiration to completely reinvent the world of physics to be excessive and presumptuous. The established parameters of classical mechanics—position, velocity, and energy—can't simply be thrown overboard. We don't live in a world made up of matrices and probabilities, he thinks. We live in our world of people, chairs, and lakes; a world where everything has its fixed place, where waves are waves, and particles are particles, and where the traditional concepts of mechanics still serve us well. There must be a connection somewhere. There must be a way to combine the jumps of the quantum world and the continuity of the classical world, somehow.

Somewhere, somehow ... Heisenberg wonders if Bohr is overcome with Scandinavian melancholy. It almost seems as if he wants to be in a state of frustration. Bohr appears to believe that quantum mechanics cannot be understood by means of classical physics, but at the same time, he wants to use classical physical concepts as the basis for quantum mechanics. How can such a contradictory stance lead to success? But Bohr seems to revel in that contradiction.

Heisenberg ponders over where the root of the problem might lie. Quantum mechanics works. Its predictions are consistently correct, down to as many decimal places as scientists care to measure. Yet, again and again, Bohr manages to find dark spots and logical ambiguities. 'Sometimes,' Heisenberg will later recall, 'I had the feeling that Bohr really was trying to lead me up the garden path. I remember sometimes getting annoyed by it.' If Bohr can lead Heisenberg up the garden path so easily, perhaps they really are on the wrong track after all.

Bohr tries to gain a footing by allowing both the wave theory and the particle theory equally. They may be contradictory theories, but they also complement each other and, taken together, offer a complete picture of the processes inside atoms.

All this interpreting goes against Heisenberg's nature. And for him, Bohr's constant talk of waves is too much of a concession to Schrödinger. In Heisenberg's quantum mechanics, there is no talk of waves. He wants to 'listen to the maths', as Paul Dirac put it. Heisenberg believes his theory is its own interpretation, without the need for any additional philosophising. For some of its parameters, matrix mechanics has already provided an interpretation: for the time-average values of energy, for example, or the electric moment, and momentum. Heisenberg hopes the process will be the same for the rest of the values; that the correct interpretation will emerge from the equations by means of 'pure logical reasoning', with no room for discussion.

Heisenberg is not pleased when Born publishes his probability interpretation in summer 1926. Born uses Schrödinger's wave mechanics to examine atomic collisions, and conjectures that an electron's wave function is a measure of the probability of encountering that electron in any given location. 'Interpretation' and 'conjecture' are not words that Heisenberg is comfortable with, but perhaps there needs to be room for interpretation after all?

Months pass, but the differences remain. In the run-up to Christmas 1926, one puzzle especially keeps Bohr and Heisenberg up talking late into the night: the puzzle of cloud-chamber tracks.

Cloud chambers have been around since the early days of atomic physics, when experimenting was still a manual craft. Niels Bohr first heard of them in 1911, when Ernest Rutherford sang the praises of the device to him. It was invented by the Scottish physicist and meteorologist C.T.R. Wilson: a sealed box with a viewing window, containing air that is saturated with water vapour. Wilson invented it with the intention of creating artificial clouds in order to study in the laboratory the light phenomena in clouds when the sun shines through them — phenomena known as coronae and glories. He expanded the air in the chamber, causing it to cool, which in turn caused the water vapour to condense on the specks of dust in the chamber, forming tiny droplets. Wilson discovered that clouds still formed in his chamber, even when he removed all the dust. Considering the cause for this, he concluded the only explanation was that the water was condensing on the ions in the air. Continuing to experiment with his chamber, Wilson found that when radiation passed through the chamber, tearing electrons from the molecules of air — that is, ionising them — they left tracks of droplets in the vapour, similar to the vapour trails that planes leave in the sky. Wilson had created a means by which physicists could observe the previously invisible trajectories of the particles given off by radioactive substances. The cloud chamber is the ancestor of the house-sized detectors in the particle accelerators of today.

In the 1920s, measuring technology is still far from being sophisticated enough to detect individual electrons. But in a cloud chamber, their tracks can be observed with the naked eye — and they really do look like the vapour trails of tiny aeroplanes. Just as classical physicists would expect.

Classical physicists, yes, but not the quantum physicists of the 1920s, like Werner Heisenberg and Erwin Schrödinger. Heisenberg, with his matrix mechanics, wants to do away with the classical particle orbits in atoms. According to Schrödinger's wave mechanics, electrons spread out through space over time. But in the cloud chamber, they appear to follow defined trajectories. How does that fit together?

'Can the natural world really behave in the absurd way atomic experiments make it appear to?' Heisenberg muses. Before he can formulate his own answer, Bohr responds with a resounding 'Yes!', and goes on to state that the key role played by observation and measurement in the interpretation of quantum mechanics frustrates any attempt to find regular patterns or causal relationships in nature.

Heisenberg doesn't like any of this. The endless discussions with Bohr, which always end up cloaked in the fog of philosophy, are draining. Often, they lead only to exasperation. 'Science arises out of conversation,' Heisenberg is famous for saying, but not like this!

Bohr and Heisenberg are still locked in endless, fruitless dispute in the first few weeks of 1927, until they reach a 'state of exhaustion', as Heisenberg later writes, which 'sometimes caused tension due to our different ways of thinking'. Each has explained his opinions to the other so many times that both feel they may as well be talking to a brick wall and are completely frustrated at the other's refusal to understand.

In February, also tired of having constantly to persist, Bohr leaves for a four-week skiing holiday in Gudbrandsdalen in Norway. The original idea was for Heisenberg to join him, but neither mentions that plan now. They both need some peace and a rest from each other.

Four weeks without Bohr. Heisenberg feels liberated. He's free to take walks round the park without them involving a discussion with Bohr. He sleeps better, and is pleased 'to be able to think about these hopelessly complicated problems alone'.

But in his mind, he can still hear Bohr's objections. What if Bohr is right, and position and velocity do still have meaning, albeit not their classical one? What could their meaning be, if particles are waves? How could that meaning be expressed in equations, rather than in Bohr's haze of words?

One evening, Heisenberg is sitting in his attic flat, where he can now think, undisturbed, about matrices and about particle trajectories. On the one hand, there are cloud-chamber tracks, which anyone can see. On the other hand, there is quantum mechanics, which does not admit classical particle trajectories. Is the theory wrong, perhaps? No, Heisenberg insists to himself, it is 'far too convincing to allow it to be changed'. The gap between theory and reality seems to be totally unbridgeable.

Heisenberg can't stop thinking on this winter evening. There must be a connection between the two sides. Around midnight, he suddenly remembers something Einstein once said to him. The previous summer, after their walk through the streets of Berlin, Einstein brought up the issue of cloud-chamber tracks as an argument against Heisenberg's theory. 'It may be heuristically useful to keep in mind what one has actually observed,' Einstein had told him at the time. 'But, on principle, it is quite wrong to try founding a theory on observable magnitudes alone. In reality, the very opposite happens. It is the theory that decides what we can observe.' Heisenberg had wondered about the question of which comes first, theory or observation. At the time, it seemed obvious to Heisenberg that observation comes first. You have to be able to see what you're thinking about. Isn't that the core principle of empirical research?

Now, as he's thinking about particle trajectories late at night, Einstein's words suddenly take on a new meaning for Heisenberg. The theory decides what we can observe. 'I was immediately convinced that the key to the gate that had been closed for so long must be sought right there,' Heisenberg would later recall.

Heisenberg is excited, but now his agitation is productive. Unable to remain seated at his desk, he rushes down the stairs and out into the neighbouring park. Walking in the clear night air, along paths lined with plane, linden, and ginkgo trees, Heisenberg goes over it all in his head once again, from the beginning. What exactly do we see when an electron shoots through the vapour in a cloud chamber? We don't see the electron actually flying. We don't see a continuous flight trajectory—that's just a construct in our head. What we actually see is a series of individual water droplets that have condensed due to the electron's passing. Each droplet is much bigger than the electron, and its position is only approximately where the electron flew by. So, what we are really seeing is not a continuous flight path, but 'a series of discrete and ill-defined spots through which the electron has passed'. We don't know how it gets from one of those spots to the next. Just as we don't know what an electron does when it jumps from one energy level to another in an atom.

What is the result when you take Einstein at his word? When you ask the theory what can be observed? In the darkness of

Fælledparken, Werner Heisenberg realises the key question is: 'Can quantum mechanics represent the fact that an electron finds itself approximately—that is, with a certain degree of inaccuracy—in a given place, and that it moves approximately—again, that is, with a certain degree of inaccuracy—with a given velocity, and can we make these approximations so close that they do not cause experimental difficulties?'

Rushing back to the institute and up the stairs to his desk, Heisenberg grabs pencil and paper, and, scribbling down some equations, works out the answer: Yes! It seems that quantum mechanics itself sets the limits on what can be measured and what cannot. But how does it set those limits, and where?

Quantum mechanics forbids a precise determination of both the position and the velocity of a particle at the same time. You can either measure its position accurately, or measure its velocity precisely, but you can't do both simultaneously. That means, if you want to know one of those values precisely, you have to enter into a bargain with nature and agree to forgo knowledge of the other. You can only look inside the atom with one of your two eyes at a time—either your position-seeing eye or your velocity-seeing eye. If you look with both eyes at the same time, the world goes fuzzy.

Heisenberg has struck upon the relationship between the inexactness of position and motion that will form the core of quantum mechanics from now on: the uncertainty principle. Expressed in the vocabulary of physics, the product of the uncertainties in position and momentum cannot be smaller than the value of Planck's constant. If Heisenberg is right, no experiment on the atomic scale can cross the limit set by the uncertainty principle. Of course, he can't strictly 'prove' it, but he's pretty sure he's hit on something, 'since the processes involved in the experiment or the observation have necessarily to satisfy the laws of quantum mechanics'.

Heisenberg has achieved what he and Bohr had failed at for months. He has established a bridge between the mathematics of quantum mechanics and cloud-chamber observations. And it was his greatest critic, Albert Einstein of all people, who gave him the crucial hint when he said, 'It is the theory that decides what we can observe.'

Over the next few days, Heisenberg investigates how sturdy his 'bridge' is. He tests the uncertainty principle in the laboratory of his mind, playing out one thought experiment after another to see if the principle can be outmanoeuvred in any way. Perhaps with a microscope strong enough to observe both the position and velocity of a moving electron at the same time? But such a microscope, with a sufficiently high resolution, would have to run on high-energy gamma rays, and those gamma rays would knock the electron off course. The formula for the resolving power of a microscope: that's the question that nearly cost him his doctorate four years earlier. Now he's using it to support the most important discovery of his career.

Heisenberg works all this out while Bohr is away skiing in Norway. He writes a fourteen-page letter to Wolfgang Pauli explaining his discovery in precise detail. Fearing Bohr's anger, he seeks Pauli's support. He sends only a short note to Bohr, saying that he has 'made progress'.

'Day is dawning in quantum mechanics,' replies Pauli. Encouraged, Heisenberg dares to revolt against his boss, expands his letter into a research paper, and sends it to *Zeitschrift für Physik* for publication, all before Bohr gets back from Norway.

On his return, Bohr reads Heisenberg's manuscript — then reads it again, first in amazement, and then with concern. When the two men meet to discuss it, Bohr tells a horrified Heisenberg that there is a fatal flaw in his reasoning. His thought experiment with the gamma-ray microscope is mistaken, Bohr says. Heisenberg sees gamma rays as streams of particles. Wrong, says Bohr: they're waves. They're particles, Heisenberg contradicts him. Waves, Bohr insists. It's the same, wretched old game.

While in Norway, Bohr has been working on his own idea about how to overcome the paradoxes of quantum mechanics. He calls the result his 'complementarity principle'. There are situations, he claims, in which we can understand one and the same event from two modes of interpretation — for example, as waves and particles. The perspectives are mutually exclusive, but also mutually complementary, and only together do they describe the event completely.

Bohr believes his idea is the better one, and considers Heisenberg's uncertainty principle to be just a special case of his principle. Heisenberg has no more appetite for arguing with him. For the next few days, he manages to avoid any discussion with Bohr. Bohr hopes that this is a sign that Heisenberg has seen the error of his ways. But Heisenberg remains resolute. Bohr urges him to retract his article. Heisenberg bursts into tears. It can be painful when a son breaks away from his father.

Bohr also realises that enough is enough; the two bury the hatchet, and agree that their theories are actually almost the same. Rather than continuing to insist that Heisenberg withdraw his article, Bohr is now satisfied for his assistant to mention his objections in an 'addendum to the errata' of his publication. At Heisenberg's request, Bohr sends the paper on his uncertainty principle to Einstein in April 1927, accompanied by a covering letter praising it. Einstein does not respond. This is the birth of the theory of quantum mechanics that will become known as the 'Copenhagen interpretation'—much to the annoyance of Max Born in Göttingen. Matrices, probabilities, uncertainty, waves, particles, complementarity—they're all included in the interpretation.

In his paper, Werner Heisenberg shakes the principle that Albert Einstein and Erwin Schrödinger consider to be the absolute foundation of physical science: the principle of causality. 'In the sharp formulation of the law of causality, "When we know the present precisely, we can predict the future," it is not the conclusion but the assumption that is false,' he writes. According to him, we cannot know the current state. We can't even know both the precise position and the precise velocity of an electron at the same time, so all we can do is calculate the probabilities of a plenitude of possible future positions and velocities of the electron. Heisenberg concludes his paper with the sentence, 'It follows that quantum mechanics establishes the final failure of causality.' Even Einstein didn't dare to go that far when he revolutionised our concepts of space and time with his theory of relativity. The clockwork universe as conceived by Newton no longer exists. And Immanuel Kant's assertion that 'All changes occur according to the law of cause and effect' is no longer applicable.

The hope 'that behind the perceived statistical world there still hides a "real" world in which causality holds' is 'fruitless and senseless', Heisenberg writes. 'Physics ought to describe only the correlation of observations.'

It's now time for Heisenberg to move on; he's learned enough in Copenhagen. Accepting the professorship in Leipzig that was offered to him a year before, he becomes the youngest full professor in Germany, at the age of twenty-five. After leaving Copenhagen, however, Heisenberg feels remorseful, and writes in a letter to Bohr, dated June 1927, that he feels ashamed 'to have given the impression of being quite ungrateful'. He adds, 'I reflect almost every day on how that came about and am ashamed that it could not have gone otherwise.' Later that year, Heisenberg visits Copenhagen again, to seek a reconciliation with Bohr.

It takes some time for other physicists to assess the significance of Heisenberg's uncertainty principle. Some theoreticians point out that Max Born had already abolished the principle of causality the year before. Some experimental physicists see the uncertainty principle as a challenge, and try to improve their instruments enough to capture a sharper image of quantum phenomena. But that's based on a misunderstanding. It's not that the world looks fuzzy; the world *is* fuzzy. 'We must accept,' says Heisenberg, 'that our words don't fit.'

COMO, 1927

Dress rehearsal

Summer 1927 on Lake Como, in the far north of Italy, close to the Swiss border. This is where the prominent theologian Romano Guardini usually spends his summer holidays. The recent rapid rise of science has him worried. Guardini describes the Modern Age as an epoch in which 'any increase in knowledge-and-technology-based power was unquestioningly taken to be a gain'. However, he says, 'the certainty of that conviction has been shaken'. Now, he believes, it is important not to increase power through more knowledge, but to tame that power. Otherwise, warns Guardini, the result will be a 'global catastrophe'.

Close by the lake, in the town of Como, a gathering is being held this summer of physicists with an untameable thirst for knowledge, including those who will be involved in developing the atomic bomb only a short time later: Werner Heisenberg and Enrico Fermi, the great talent of Italian physics. Niels Bohr is there, still wrestling with interpretations of quantum mechanics. Wolfgang Pauli is present, as well as Max Born, Hendrik Lorentz, Arnold Sommerfeld, Louis de Broglie, Max Planck, Arthur Compton, and John von Neumann. According to a report in the *Physikalische Zeitung* newspaper, physicists from fourteen countries are gathered in Como: Switzerland, Sweden, the United States of America, Spain, Russia, the Netherlands, Italy, Great Britain, India, Germany, France, Denmark, Canada, and Austria.

The congress marks the centenary of the death of Alessandro Volta, the inventor of the battery, who also gave his name to the electrical unit the volt. The meeting is part of the events around the 'Volta Centenary Exhibition', with which the government of the *Duce del*

Fascismo, Benito Mussolini, is trying to promote the natural scientist as a national hero.

Albert Einstein is not present, having refused to set foot in fascist Italy. Erwin Schrödinger has also cancelled. Just a couple of weeks earlier, he had become Max Planck's successor, and is busy with his move to Berlin. Bohr will have to wait another couple of months before the two meet in Brussels.

It's been six months since Werner Heisenberg wrote his groundbreaking paper on the uncertainty principle. Since then, Niels Bohr has been working on his own paper. He both writes and has others write for him. After Heisenberg fled from Copenhagen to Leipzig, exhausted by all the discussions with him, Bohr's new assistant, Oskar Klein, now has to put up with his daily ramblings. Bohr thinks aloud about the uncertainty principle, slowly and deliberately, struggling to find a way to formulate his thoughts, trying sentences for size. In the evenings, Klein writes up whatever he's been able to glean from Bohr. In the mornings, Bohr throws Klein's notes away and starts all over again. When the Bohrs leave Copenhagen to spend the summer at their country house on the coast, north of the city, Klein has no choice but to go with them, and his travails continue. Even Margrethe Bohr, usually the epitome of calm and cheerfulness, is reaching the end of her tether, and sometimes even bursts into tears—not, like Heisenberg, out of opposition to her husband's scientific position, but because his mind is constantly elsewhere. This was supposed to be a family holiday, but Niels constantly leaves Margrethe alone with their five children.

Niels Bohr delivers his talk on 16 September 1927 in the grand wood-panelled hall of the Istituto Carducci. Before then, he has corrected his manuscript over and over again. He has fiddled with it until the very last minute, fine-tuning it, and his notes are a mess of deletions, additions, marginal remarks, and arrows when he steps up before his audience. He looks up briefly and collects his thoughts before launching his talk in heavily Danish-accented English, mumbling so quietly that many in the audience involuntarily lean forward, straining to hear. Those sitting at the back have little chance of catching every word. Bohr mentions his concept of complementarity publicly for the first time. He then goes on to explain Heisenberg's uncertainty principle, and finally describes the role that measurement plays in the interpretation of quantum mechanics. Any of those topics would easily have been enough to fill an entire lecture. And each is a challenge, even for such an expert audience as this. It's also far from obvious what each topic has to do with any of the others. Bohr skilfully connects them, also weaving Schrödinger's wave function and Born's probability interpretation into the fabric. He tries to pull off a great coup: to introduce a new physical understanding of quantum mechanics. The audience is both confused and impressed. This is the day that the phrase 'the Copenhagen interpretation', that most elegantly ordered miscellany of ideas and formalisms, enters into the vocabulary of physical scientists.

Bohr's lecture is the distillation of the months of wrestling with the interpretation of quantum mechanics that he engaged in with Heisenberg. It begins with the idea that every measurement interferes with the system being measured. So far, so good. Bohr then moves on to the idea that, in quantum mechanics, it is the measurement that defines the measured value. The result of a measurement depends on what is being measured. That's well and good. Now, however, Heisenberg has shown that measuring one value makes it impossible to measure another at the same time. So, while such a measurement increases knowledge of the system being measured in one respect, it reduces knowledge in another. If we know the position of a particle, we cannot know its momentum; and, what is more, Bohr continues in his tortured and tortuous sentences, it *has* no momentum until we measure it, and when we do, its position becomes unknowable.

To simplify matters, he introduces the concept of complementarity. A quantum system can only be understood in terms of opposites. Waves or particles. Position or momentum. They are contradictory, but also complementary. That's complicated. Bohr confronts his audience with sentences such as: 'The very nature of the quantum theory thus forces us to regard the space-time coordination and the claim of causality, the union of which characterises the classical theories, as complementary but exclusive features of the description, symbolising the idealisation of observation and definition respectively.'

Complementarity is the answer; it attempts to make the incompatible compatible, to combine Heisenberg's uncertainty principle, Born's probabilities, and Schrödinger's waves, which are nothing like the classical waves Schrödinger claims they are, because, Bohr says, they vibrate in predictable ways only when they are not being measured. But, most importantly, Bohr wants complementarity to render all of the above compatible with his own correspondence principle, the foundation of his quantum theory. Each practical description of the behaviour and properties of a quantum system must ultimately be able to be converted into the language of classical physics. We can't observe probability clouds. We can't measure blur. Experiments produce real results.

What exactly does Bohr mean by all this? No one understands it completely, not even Bohr himself. Others suspect that Bohr is trying to tell them something they already know, but in a completely incomprehensible way.

At the end of Bohr's talk, Max Born rises from his seat to briefly express his agreement. He says the theory is coherent and can be used to make calculations and predictions, and that's enough for him. Then Werner Heisenberg gets up to speak. Only a few insiders know of the battles that the inventor of quantum mechanics has fought with his mentor Bohr over the previous months. Those differences now seem to be consigned to the past. Heisenberg is no longer fighting Bohr, and has nothing but thanks and praise for him.

No one contradicts Bohr. Those who might have, Einstein and Schrödinger, are not in attendance. So Bohr's interpretation of quantum mechanics begins to take hold, due to a lack of opposition.

Bohr draws up a written version of his lecture for the English-language science journal *Nature*, and, once again, it's a tortuous process, taking him several months to complete. He drafts, discards, and reworks his text, with the help of Wolfgang Pauli. The publishers of *Nature* send enquiries and entreaties. Bohr apologises for the delay. But he insists on his sentences, many of which push the bounds of

English grammar: 'Indeed, we find ourselves here on the very path taken by Einstein of adapting our modes of perception borrowed from the sensations to the gradually deepening knowledge of the laws of Nature.' And that's how the article goes to print. The publishers produce a commentary on Bohr's text in which they express their hope that his conception is not the last word on quantum mechanics, as it 'obviously cannot be clothed in clear language'.

Como is actually just the dress rehearsal for a great confrontation that is still to come. Only a few weeks after the physicists disperse from Como, they will congregate in Brussels to discuss 'electrons and photons' at the fifth Solvay Conference. This time, Albert Einstein and Erwin Schrödinger will be there.

BRUSSELS, 1927

The great debate

The Dutch physicist Hendrik Antoon Lorentz is a fine man, well liked by everybody for his friendly and thoughtful manner. He speaks fluent English, German, and French. Einstein once described Lorentz as 'a wonder of intelligence and fine tact, a living work of art!' On 2 April 1926, Lorentz attends a private audience with the Belgian king, Albert I. His mission is politically sensitive: he wants the king's permission to invite German physicists to come to Belgium.

Albert I, known as the 'Soldier King', is not exactly well disposed towards Germans. The First World War began with Germany's invasion of neutral Belgium in 1914. As commander of the 'Army Group Flanders', Albert led the final offensive against German troops that ended with the liberation of western Belgium and the armistice of 11 November 1918. He has no intention of allowing even a single German to set foot on Belgian soil again.

But with his tactful and disarming nature, Lorentz wins Albert's permission to invite German physicists to Brussels for the fifth Solvay Conference, planned for the autumn of the following year. As chairman of the conference's scientific committee, Lorentz manages to persuade King Albert that now, seven years after the end of the war, it's time to take a more conciliatory stance towards Germans and to work towards a better understanding between nations. Lorentz argues that science can play an important part in that process, also pointing out that ostracising German scientists would be difficult over the long term, in view of their huge contribution to physics. Before the day of his royal audience is over,

Lorentz writes to Einstein to notify him of his breakthrough.

It's a spectacular change of heart; German scientists have indeed been sidelined by many countries since the war. They are isolated from the international scientific community as pariahs. Albert Einstein was the only German invited to the third Solvay Conference, in 1921, and he wasn't completely German anyway. However, Einstein had decided not to take part in that event in protest at the exclusion of other German scientists. Instead, he had toured the US, giving lectures and raising funds for the establishment of the Hebrew University of Jerusalem. Two years later, he announced he would also decline the invitation to the fourth Solvay Conference due to the continuing boycott of German scientists. 'No politics should be carried into scientific endeavours,' Einstein wrote to Lorentz at the time, 'and one should generally not make individual people responsible for the state to which they happen to belong.'

Now Hendrik Lorentz manages to persuade the Belgian king that reconciliation is the best way forward, and that Germans can be invited to this year's conference. Even Lorentz himself is surprised. He's been president of the International Committee on Intellectual Cooperation, an organisation of the League of Nations, since taking over from Henri Bergson in 1925. Until now, Lorentz has been pessimistic about the prospect of German scientists being allowed to attend international conferences again.

The international mood is changing. In October 1925, the Locarno Treaties were negotiated in an elegant palazzo on Lake Maggiore, after months of preparation by German, French, and Belgian diplomats. They agreed on recognised borders between their countries and on an arbitration system for disputes. The treaties paved the way for Germany's acceptance into the League of Nations a year later. Germany is gradually recovering from the aftermath of World War I. Post-war depression is giving way to excitement about new ideas, art forms, and technological developments. People marvel at the wonders of aviation, drive around in automobiles, speak to each other by telephone, and flock to the cinema.

After years of nationalism, a new, international spirit grips the

Solvay Conference, Brussels, 1927

Standing, from left to right: Auguste Piccard, Émile Henriot, Paul Ehrenfest, Édouard Herzen, Théophile de Donder, Erwin Schrödinger, Jules-Émile Verschaffelt, Wolfgang Pauli, Werner Heisenberg, Ralph Howard Fowler, Léon Brillouin

Sitting, second row, from left to right: Peter Debye, Martin Knudsen, William Lawrence Bragg, Hendrik Anthony Kramers, Paul Dirac, Arthur Holly Compton, Louis-Victor de Broglie, Max Born, Niels Bohr

Sitting, front row, from left to right: Irving Langmuir, Max Planck, Marie Curie, Hendrik Antoon Lorentz, Albert Einstein, Paul Langevin, Charles-Eugene Guye, Charles Thomson, Rees Wilson, Owen Willans Richardson world. In May 1927, a young American by the name of Charles Lindbergh becomes the most famous man in the world when he flies his metal-tubing, wood, and treated-canvass monoplane, the *Spirit of St. Louis*, from New York to Paris. With 1,700 litres of fuel in the tank, the aircraft carries even less luggage than Heisenberg took on his trip to Heligoland. In fact, all Lindbergh packs is five sandwiches, pragmatically pointing out, 'If I get to Paris, I won't need any more, and if I don't get to Paris, I won't need any more either.' Lindbergh is two months younger than Heisenberg.

It is in the new spirit of optimism that Albert I is persuaded to reopen his country to German scientists. With Einstein's help, Lorentz starts planning the fifth Solvay Conference. Its official title is 'Electrons and Photons', but unofficially the event is dominated by quantum mechanics. Barely six months after Heisenberg published his uncertainty principle, scientists are discussing its significance for the future of physics. The fifth Solvay Conference will become the most famous meeting in the history of physics, and scientists will still be discussing it nearly a hundred years later, especially as the Brussels event is seen as beginning of the duel between the world's top two physicists of the time—Niels Bohr and Albert Einstein. 'No more profound intellectual debate has ever been conducted,' the English scientist and novelist C.P. Snow would later claim.

The fifth Solvay Conference takes place from 24 to 29 October 1927 in Brussels, home of the Solvay chemical company, which made its money manufacturing sodium carbonate for washing powder. Ernest Solvay, who founded the company with his brother Alfred in 1863, developed the 'Solvay process' for the production of washing soda, and is a generous patron of social and educational projects. He initiated the recurring Solvay Convention in 1911 as a select gathering of invited participants, and the guest list for the fifth convention is the most illustrious yet. Seventeen of the twenty-nine participants are past or future Nobel laureates. The only woman to take part even has two Nobel Prizes: Marie Curie. Her former lover Paul Langevin, with whom she had a scandalous affair, is also present.

The Brussels event is attended by practically every physicist who has something to say about quantum mechanics: Max Planck, Albert Einstein, Paul Ehrenfest, Max Born, Niels Bohr, Erwin Schrödinger, Louis de Broglie, Hendrik Kramers, Wolfgang Pauli, Werner Heisenberg, and Paul Dirac. This is the only time they will all be gathered together. Only Arnold Sommerfeld is missing. He had spoken out in favour of Germany's occupation of Belgium during the war, and was not invited.

For Niels Bohr, like Albert Einstein, this is the first Solvay Conference he has taken part in. Bohr missed the 1921 meeting due to illness, and he turned down the invitation in 1924, fearing that his attendance could be interpreted as tacit approval of the policy of excluding German scientists.

Expectations are high as the world's leading minds in quantum physics assemble at the Physiological Institute in Brussels' Leopold Park on the cloudy, grey morning of Monday 24 October 1927.

Hendrik Lorentz begins by welcoming the participants, and then hands over to Lawrence Bragg, the man who succeeded Ernest Rutherford as professor of physics at Manchester University. Bragg is a shy, nervous, thirty-seven-year-old Australian who won a Nobel Prize, together with his father, twenty-five years earlier for the analysis of crystal structures by means of X-rays. In his thick Australian accent, he delivers a talk on how the latest data from such analyses can shed light on the structure of the atom—all very quiet, correct, and ... a little bit boring. There follows a short debate on this rather dry subject. Heisenberg, Dirac, Born, and de Broglie have a few comments and questions. A fluent speaker of German, English, and French, Lorentz translates for the less linguistically gifted delegates. Then they all head off to lunch.

In the afternoon, the thirty-five-year-old American physicist Arthur Compton reports on his experiments with electrons and X-rays. Compton is an experimental physicist, but not just a lab rat. He's also interested in philosophical issues, such as the question of free will, while, on the practical side, he will later invent a new, gentler version of the traffic-control feature we now know as 'speed bumps'. He was awarded the Nobel Prize just weeks before the conference began, but modesty prevents him from using the now-common term 'Compton scattering' to refer to the reduction in the frequency of X-rays when they scatter particles such as electrons.

Bragg's and Compton's talks have the same underlying message. Electromagnetic theory, formulated by James Clerk Maxwell in the nineteenth century and seen as written in stone ever since, is starting to crumble. The phenomena observed by Bragg and Compton cannot be explained by electromagnetism. Where it fails, Einstein's concept of light quanta provides a way to reconcile theory and experimental results. But by the end of that first day, Einstein is the only leading physicist who has not said a word. His moment to take the floor is yet to come.

After some hesitation, Einstein had turned down an invitation to set out his position on quantum mechanics in a lecture at the conference, writing to Lorentz that he was 'not competent', and that he felt left behind by the 'tempestuous developments in quantum physics', and was not involved to the extent necessary for a substantial contribution, adding, 'I have now given up that hope.' But Einstein is not telling the whole truth here. He absolutely does want to be involved, but he holds his tongue and listens, waiting for his opportunity.

The other great eminence of quantum physics, Niels Bohr, also delivers no talk at the 1927 Brussels event. He has not been one of the driving forces behind the latest developments in quantum mechanics, but he has his protégés. Heisenberg, Pauli, and Dirac continue to work on the theory.

However, the theory itself is not really the issue at the conference; rather, it is how to interpret it. What kind of a world does it describe? How does it deal with the principle of cause and effect? Is the moon still there when no one is looking at it? Such questions used to be the business of philosophy. Now, physicists are having to consider them, too, and to find answers, if they are to understand their own theory. Bohr is convinced he can provide those answers.

Now he wants to convince Einstein. Bohr is curious: how will Einstein react to the latest developments in quantum physics?

Einstein's judgement is very important to Bohr; after all, Einstein is still considered the pope of physics.

As the conference progresses, the line of conflict between the participants becomes increasingly distinct: the old guard versus the proponents of the new quantum physics. Albert Einstein, Erwin Schrödinger, Max Planck, Hendrik Lorentz—the old guard—are defenders of the established order of classical physics, with its gently rolling waves and particles that stick to the orbits they're on. These are the 'realists', whose aim is to describe the world as it really is.

The young 'instrumentalists'—notably, Werner Heisenberg, Wolfgang Pauli, and Paul Dirac—are keen to advance quantum mechanics and apply it to the quest to find answers to the still-unanswered questions about the nature of the atom and radiation. They have no patience with anything that remotely smacks of philosophy, semantics, or nitpicking.

Only Niels Bohr refuses to be drawn into either of those camps. The wild young things are his students. But his respect for his old friend, and his own propensity for philosophical doubt, mean he can't just dismiss Einstein's objections out of hand.

The next day begins sedately, with a reception at the Free University of Brussels. But things get serious after the lunch break. The aristocratic Louis de Broglie delivers a lecture on 'The new dynamics of quanta'. Speaking in French, he tells the audience how he came upon the idea that all matter is made up of waves, and how Erwin Schrödinger developed this approach into his wave mechanics. De Broglie goes on to say that, after further consideration, he can now present his elaboration of that idea, which takes account of the current state of quantum mechanics. Clever use of Schrödinger's equation has allowed him to create a new picture of quantum mechanics without changing the maths. He rejects Bohr's mysterious talk of 'complementarity', which sees waves and particles as opposed, incomplete, mutually complementary representations of quanta. Instead, de Broglie envisions a quantum world in which waves and particles peacefully coexist.

De Broglie has attempted something that no one before him has tried: to bridge the gap between supporters of wave mechanics and the Copenhagen camp. He cautiously accepts Born's probability interpretation, pointing out that the task at hand now is to investigate how it all fits together. De Broglie sketches out a theory according to which particles are 'guided' or 'piloted', as he puts it, by waves. This allows him to keep waves in his theory while restoring particles to their rightful place, even if that rightful place now turns out to be a 'hidden variable'—hidden, that is, from quantum mechanics. Contrary to the Copenhagen interpretation, an electron does not behave like either a wave or a particle. In fact, it is both at the same time: a particle that glides atop a wave like a surfer. '*Théorie de l'onde pilote*' is de Broglie's name for his new approach; the 'pilot wave theory' in English. De Broglie's particles behave completely deterministically, but still fulfil Heisenberg's uncertainty principle because their paths are hidden from view. No measurement can capture their movements completely, just as Heisenberg's principle stipulates. It almost looks like a magic trick: the principles of causality and determinism, the two ancient pillars of physical theory, both remain valid, but so does quantum mechanics and its striking correlation with experimental results.

But the two camps are reluctant to be reconciled. Erwin Schrödinger, who doesn't believe in particles, doesn't even listen to the lecture. Wolfgang Pauli calls de Broglie's approach 'very interesting, but wrong', and goes on to say that the Frenchman's idea of wave-riding particles contradicts the current quantum theory of particle collisions. Although Pauli's attack is based on a skewed analogy, it throws de Broglie off guard, and he can only stammer in reply. Despite the fact that he's right, de Broglie stands there as if refuted, and Pauli, who isn't right, triumphantly retakes his seat.

A weightier objection comes from Hendrik Kramers. He points out that a mirror reflecting a light particle will experience a slight recoil force from the impact of the particle. Kramers says de Broglie's theory is unable to explain that recoil; now even more unsure of himself, de Broglie has no answer.

'The indeterminist school, whose followers were mostly young and uncompromising, met my theory with cold rejection,' de Broglie would later recall. Where are the equations, they ask. De Broglie has no formulae to write on the chalkboard. All he has is his idea. He hopes to gain the support of Albert Einstein. Perhaps he can help attract the undecided over to the right side? But Einstein continues to keep his silence. De Broglie is demoralised. Rather than bringing about reconciliation, he has only succeeded in turning all his colleagues against him.

On the third day, Wednesday 26 October 1927, it's the turn of the wild young things to take the floor. Born and Heisenberg deliver a joint lecture presenting their matrix-based formulation of quantum mechanics, in which random quantum jumps play a major part. Even their opening remark is a slap in the face for Schrödinger: 'Quantum mechanics is based on the intuition that the essential difference between atomic physics and classical physics is the occurrence of discontinuities.' Then comes the diplomatically important nod to their eminent colleagues sitting only a couple of metres away, as Born and Heisenberg point out that quantum mechanics is essentially 'a direct continuation of the quantum theory founded by Planck, Einstein, and Bohr'.

After describing their matrix mechanics, the Dirac-Jordan transformation theory, and Born's probability interpretation, Born and Heisenberg turn to the uncertainty principle and the 'actual meaning of Planck's constant *h*'. They maintain that this *h* is nothing less than the 'universal measure of the indeterminacy that enters the laws of nature through the dualism of waves and corpuscles'. Thus, if there were no wave-particle duality of matter and radiation, there would be no Planck's constant and no quantum mechanics. Their reasoning is a provocation for Einstein and Schrödinger: 'We consider quantum mechanics to be a closed theory, whose fundamental physical and mathematical assumptions are no longer susceptible to any modification.' In short, quantum physics is done, a fully mature theory that requires no more tinkering with, shaking up, or interpreting.

Quantum mechanics is 'closed'—that's the message Bohr, Born, Heisenberg, and Pauli have come to Brussels to spread. They're convinced they now have a coherent formulation of quantum mechanics. Just a couple of years earlier, quantum physics was still a difficult construct with a constant stream of provisional models, each soon disproved by experimental results. Now, all the foundations are united: matrix mechanics, Schrödinger's equation, and Heisenberg's recent discovery of the keystone of quantum mechanics, the uncertainty principle.

The audience is impressed by these young Germans, and sentiment now seems to be on their side, although some in the lecture hall still have their doubts, first and foremost Albert Einstein. In Einstein's opinion, quantum mechanics represents an impressive intellectual achievement, but it is not a true theory about the reality of the smallest building blocks of nature. However, in the ensuing discussion, he continues to keep his silence. No one contradicts Born and Heisenberg. Dirac, Lorentz, and Bohr comment on just a few minor details.

Why is Einstein so silent? Perhaps it's just his unshakeable equanimity. Maybe incredulous amazement? Paul Ehrenfest is determined to find out, so he passes a scribbled note to Einstein: 'Don't laugh! There's a special section in purgatory for "professors of quantum theory", where they will be obliged to listen to lectures on classical physics for ten hours every day.' 'I laugh only at their naivety,' Einstein replies. 'Who knows who will have the last laugh in a few years?'

However, Einstein is not prepared just to wait till then. He pursues a guerrilla tactic: avoiding open confrontation, but needling his opponents at mealtimes. Even early in the morning, in the breakfast room of the Hotel Metropole, he's ready with a seeming refutation of quantum mechanics. The discussion starts over coffee and croissants. Pauli and Heisenberg listen with only half an ear, dismissing the arguments, 'Yes, yes, that's right, that's right,' but Bohr listens more intently, and, after pondering throughout the morning, discusses his thoughts with Heisenberg and Pauli over lunch. By dinnertime, he has come up with a refutation of the refutation. The next day, the whole game starts over again.

Paul Ehrenfest, who is friends with both men, writes in a letter shortly after the conference is over, 'It was delightful for me to be present during the conversations between Bohr and Einstein. Like a game of chess. Einstein all the time with new examples. In a certain sense a *perpetuum mobile* of the second kind to break the uncertainty relation. Bohr from out of philosophical smoke clouds constantly searching for the tools to crush one example after the other. Einstein like a jack-in-the-box: jumping out fresh every morning. Oh, that was priceless. But I am almost without reservation pro Bohr and contra Einstein. His attitude to Bohr is now exactly like the attitude of the defenders of absolute simultaneity towards him.' After a few days, Ehrenfest confronts his friend directly. 'Einstein, I am ashamed for you, because you argue against the new quantum theory precisely as your opponents argued against the theory of relativity.' Heisenberg throws Pauli a look: at last someone has come out and said it. It's a blow for Einstein. His faithful friend Ehrenfest has switched sides.

Every night, at one o'clock in the morning, Bohr visits Ehrenfest in his room, utters a few mysterious words, and fills the room with tobacco smoke. He stays till about three, then leaves. Ehrenfest is not pleased with this behaviour: he's not a fan of nebulous words, or of stinking clouds of pipe smoke. He complains of the 'awfully Bohrian incantation terminology. Impossible for anyone to summarise.'

Erwin Schrödinger has the chance to strike back on Wednesday afternoon, but doesn't take it. Instead, he delivers a limp defence of his wave mechanics. 'Under this name at present two theories are being carried on,' he reports in English, 'which are indeed closely related but not identical.' There is really only one theory, but it is effectively split in two, he goes on. One part concerns waves in ordinary, everyday three-dimensional space, like a sound wave or a classic light wave, while the other requires a highly abstract multi-dimensional space. The problem, Schrödinger explains, is that for anything other than a moving electron, this is a wave that exists in a space with more than three dimensions. Whereas the single electron of a hydrogen atom can be accommodated in a three-dimensional space, helium, with its two electrons, needs six dimensions. Schrödinger admits himself that such ideas might curb enthusiasm for his theory. But he explains that this multi-dimensional space, known as configuration space, is only a mathematical tool, and whatever is being described, be it electrons colliding or orbiting the nucleus of an atom, the process takes place in space and time as we know them. 'In truth, however, a complete unification of the two conceptions has not yet been achieved,' Schrödinger has to admit.

Schrödinger's hope that such a unification might still be achieved is not shared by the other major theoreticians present. Although most physicists prefer to use his wave mechanics over Heisenberg's matrices in their calculations, very few believe that wave theory provides a realistic picture of actual charge-carrying electron clouds and mass distributions. They accept Born's probability interpretation, which Schrödinger steadfastly rejects. He makes no bones about his disdain for the idea of 'quantum jumping'.

From the moment Erwin Schrödinger received his invitation to speak in Brussels, he knew there would be a clash with the proponents of matrix mechanics. So he's prepared for the attacks that begin during the post-lecture discussion. But he's not prepared for how vehement they are. The first stab comes from Bohr, who asks Schrödinger whether the 'difficulties' he mentioned late in his lecture imply that the results mentioned early in the talk are actually wrong. Schrödinger answers confidently. Then Born pipes up, challenging another of Schrödinger's calculations. Now starting to lose his patience, Schrödinger assures him that his calculations are 'completely correct and rigorous, and this objection by Mr Born is completely groundless'.

After a couple of others have spoken, it's Heisenberg's turn: 'Mr Schrödinger says at the end of his report that the discussion he has given reinforces the hope that when our knowledge is deeper it will be possible to explain and to understand in three dimensions the results provided by the multi-dimensional theory. I see nothing in Mr Schrödinger's calculations that would justify that hope.' Schrödinger argues that his 'hope of achieving a three-dimensional conception is not quite utopian'. A few minutes later, the discussion ends. Offended, Schrödinger never speaks another word at the conference.

The Thursday is a free day. Lorentz, Einstein, Bohr, Born, Pauli, Heisenberg, and de Broglie take the train to Paris to attend the ceremonies at the Académie des Sciences to mark the 100th anniversary of the death of Augustin Fresnel, one of the pioneers of wave optics. 'The real battle begins tomorrow,' writes Werner Heisenberg on the evening of Thursday 27 October, in a letter to his parents.

Heisenberg is enjoying his stay in Brussels: the smoking and drinking together with his colleagues, the Hotel Metropole, the Opera, which is 'really decent', and, in particular, his membership in the international scientific community. And everyone talking about his theory, the uncertainty principle.

On Friday, the discussions become more heated. After the official talks are over, the program moves on to a 'general discussion of the ideas proposed'. Lorentz is bombarded from all sides by voices speaking in English, French, and German, all trying to be heard. Ehrenfest jumps up and writes on the chalkboard, 'The Lord did there confound the languages of all the Earth'—a quote from the biblical story of the Tower of Babel in the Book of Genesis. Laughter accompanies him as he returns to his seat.

Lorentz tries to steer the discussion towards the issues of causality, determinism, and probability. Do quantum events have a cause? That's the question Marie Curie had wrestled with a quarter of a century earlier. As Lorentz puts it, 'Could one not maintain determinism by making it an article of faith? Must one necessarily elevate indeterminism to a principle?' Offering no answer of his own, Lorentz invites Bohr to share his thoughts on the 'epistemological problems confronting us in quantum physics'. This is the moment of truth, and everyone present knows it. This is when Bohr will attempt to convince Einstein of the correctness of the Copenhagen interpretation.

Bohr approaches the lectern. He addresses the entire room, but really he's speaking to Einstein. He recalls the lecture he gave in Como, which was directed at Einstein, but which Einstein had not heard because he was not present at the gathering. Bohr outlines his belief that wave-particle duality is an intrinsic feature of nature that is explicable only within the framework of complementarity as developed by him, and that complementarity underpins the uncertainty principle, which exposes the limits of applicability of classical concepts. However, the ability to communicate unambiguously the results of experiments probing the quantum world, Bohr explains, requires the experimental set-up as well as the observations themselves to be expressed in a language 'suitably refined by the vocabulary of classical physics'.

Einstein listens carefully. Eight months earlier, in February 1927, as Bohr was pondering his principle of complementarity, Einstein gave a lecture in Berlin on the 'Theoretical and experimental aspects of the question of the generation of light'. He argued that instead of either a quantum or a wave theory of light, what was needed was a synthesis of both conceptions. This was a synthesis he had already called for in 1905, when he invented the concept of light quanta, but which had not yet come about. And now here is Bohr imposing a segregation of the two through complementarity. It's either waves or particles, depending on the choice of experiment. Never both. No synthesis.

The issue is greater than just waves versus particles. It's about the very nature of physics itself. Scientists have always seen themselves as passive observers of nature. Yes, they intervene when they conduct experiments, but not when they simply observe. There has always been a sharp distinction between the observer and observed. In classical theory, the observer has no influence on the observed, and Einstein wants to keep it that way.

However, Niels Bohr does not. He claims that in the world at the atomic level, such a separation between the observer and the observed does not exist. For that world, according to the Copenhagen interpretation, Bohr identifies what he calls the 'essence' of the new physics — the 'quantum postulate'. When investigating atomic phenomena, the interaction between what is measured and the measuring equipment means, according to Bohr, that 'an independent reality in the ordinary physical sense can neither be ascribed to the phenomenon nor to the agencies of observation'.

Reality, as Bohr envisions it, does not exist when not under observation. An electron simply does not exist at any place or move with any speed until an observation or measurement is made of it. In between measurements it is meaningless to ask what the position or velocity of an electron is. An unobserved electron does not exist. Only in the act of measurement does the electron become real. It is wrong to think that the task of physics is to find out how nature is, Bohr reasons. Physics concerns only what we can *say* about nature.

Einstein takes a different view: 'What we call science has the sole purpose of determining what *is*.' The aim of physics is to explain reality objectively, independently of any particular point of view. This is the chasm separating the beliefs of the Copenhagen set and Einstein. 'Atoms or the elementary particles themselves are not real,' says Heisenberg, 'they form a world of potentialities or possibilities rather than one of things or facts.' For Bohr and Heisenberg, the transition from the 'possible' to the 'actual' takes place during the act of observation. For Einstein, that assertion no longer belongs in the realm of science. Science doesn't invent nature; science examines nature. The very soul of physics lies at the core of this dispute.

After Bohr's contribution, Einstein still keeps his silence. Three audience members raise their hands to comment. Then Einstein indicates to Lorentz that the moment has arrived. A hush descends over the audience. All eyes are on Einstein as he rises from his seat and strides up to the chalkboard. With the tips of his starched collar neatly folded forward and his long necktie, he looks like a figure from the previous century.

Einstein makes a cautious start: 'Conscious as I am of the fact that I have not entered deeply enough into the essence of quantum mechanics, I want nevertheless to present here some general remarks.' This is, of course, false modesty: 'I've thought a hundred times more about quantum problems than about the general theory of relativity,' he later admits to a friend. Some people believe Einstein has failed to understand quantum mechanics. But nothing could be farther from the truth; he understands it better than anyone else. He just doesn't agree with it, considering it incomplete.

Einstein doesn't respond to Bohr's contribution, which was so unmistakably directed at him. He ignores Bohr's attempt to win him over to quantum mechanics, and says not a single word about Bohr's analysis of the wave-particle puzzle, or about the idea of complementarity, or about any of Bohr's philosophical expositions.

He takes immediate aim at Bohr's weakest point: the claim that quantum mechanics 'exhausts the possibilities of accounting for observable phenomena'. Why should this ugly theory be the one to draw the line for the limit of what can be examined? Einstein sets about proving that quantum mechanics, as Bohr understands it, is not a complete theory free of contradiction. And he employs his favourite tool to do so: the thought experiment.

'Imagine an electron striking a screen,' Einstein begins, approaching the board and grabbing a piece of chalk. Drawing vertical lines to represent the path of the electron, he adds a horizontal one to represent the screen, lifting his chalk briefly to leave a gap to represent a small slit in the screen. The slit will scatter the electron as it passes through, he explains, with his back half-turned to the audience. Beyond the screen, he draws semicircles to represent the Schrödinger waves of the electron emanating from the slit towards a photographic plate. The electron will strike the plate. If we observe the electron here — Einstein indicates a point near the top of the plate — it cannot possibly strike the plate here — he points to a location near the bottom of the plate. But the Schrödinger wave, which, according to the Copenhagen interpretation, expresses 'the probability that this particle is located at a specific position', strikes the plate across its full width, not just at a single point. It has 'no preferred direction', Einstein continues. As soon as the electron is observed at some point on the plate, the wave must suddenly disappear from everywhere else on the plate: the infamous 'collapse of the wave function'. That cannot be the case, Einstein explains, because it would be a violation of the requirements of relativity, which forbid such instantaneous (that is, faster-than-light) remote effects. If an event at one point in space is the cause of another at a different point in space, then there must be a time lapse between them to allow a signal to travel at the speed of light from one location to the other. Therefore, Einstein reasons, Bohr's interpretation of quantum mechanics is unable to provide an explanation for real-world events.

Einstein proposes a different explanation. Each electron remains a particle and follows one of many possible trajectories through the slit until it hits the photographic plate. The spherical waves he described before still exist; however, they do not correspond to individual electrons, but to 'a cloud of electrons'. Quantum mechanics does not provide any information about individual processes, only about what he calls an 'ensemble' of processes. Einstein counters the Copenhagen interpretation with a 'purely statistical' interpretation.

Putting down the chalk and brushing the dust from his hands, Einstein concludes, 'In my opinion, the only way to avoid this objection is to describe the process not only by means of a Schrödinger wave, but also by localising the particle as it propagates. I think de Broglie is right to look in that direction.' In other words, quantum mechanics may not be wrong, but it is incomplete. The reality of quantum processes lies deeper. De Broglie is on the right track, and Heisenberg and Pauli are barking up the wrong tree. Einstein resumes his seat.

Bohr, Heisenberg, Pauli, and Born look at each other agog. That was supposed to be Einstein's rebuttal of quantum mechanics? True, the wave function collapses suddenly, but it is an abstract probability wave, not a real wave propagating through the three-dimensional space we live in.

'I feel myself in a very difficult position because I don't understand what precisely is the point which Einstein wants to make,' says Bohr. 'No doubt it is my fault.' He follows up with an astonishing remark: 'I do not know what quantum mechanics is. I think we are dealing with some mathematical methods that are adequate for a description of our experiments.'

Instead of responding to Einstein's analysis, Bohr simply restates his own views. He starts talking about wave-particle duality and complementarity once again. He speaks of observations, while Einstein speaks of reality. In their first public dispute, the two old masters of quantum physics are talking completely at cross-purposes. 'The confusion of ideas reached its high point,' comments Paul Langevin.

After presenting his arguments, Einstein falls silent again. The discussion draws to a close, and the participants leave the Physiological Institute. But the dispute has only just begun. When Niels Bohr and Albert Einstein meet again in the art deco hall of the Hotel Metropole one evening, de Broglie reports seeing the two rivals engrossed in their 'duel', but he's unable to follow their conversation as he doesn't speak German.

Einstein maintains that a fundamental physical theory cannot possibly be a purely statistical one. Of course, statistical theories do exist in physics—for example, in thermodynamics or statistical mechanics. But they are not fundamental theories. Statistics is just a way to fill in gaps in knowledge of the processes being described. And the same is the case in quantum mechanics, Einstein remarks. Again and again, he repeats the phrase 'God doesn't play dice'.

'Einstein,' retorts Bohr at some point, 'it can't be our place to tell God how to rule the world.' However, Bohr does not succeed in undermining Einstein's objection based on the collapse of the wave function, soon to become known as 'the measurement problem'. How can it be that an unobserved electron can flow through space as a wave, only to collapse, as if by magic, into a single point the moment it is measured? Bohr does not have an answer to this question. Quantum mechanics does not provide an answer.

Einstein tries to follow the path of the electron from the slit in the screen to the photographic plate. Perhaps the uncertainty principle can be tricked by placing another screen, this time with two slits, between the first one and the plate? Bohr ponders this for a couple of hours. Has Einstein taken into consideration the accuracy with which the position of the slit in the screen is known? And what about the recoil from the electron as it passes through the slit? Over and over again, before their mind's eye, Bohr and Einstein watch the electron pass through the apparatus. Bohr is hardly able to formulate a full sentence. Einstein continues to hope. Until Bohr shows that the whole apparatus that is meant to reveal the position of the electron is subject to the uncertainty principle itself, which leads to the position being uncertain again.

This worries Einstein. He doesn't have a ready riposte. By the next morning, he has come up with a more sophisticated thought experiment. He adds a new screen with more slits, and more measuring instruments. As the experiment gets increasingly complicated, he gets increasingly irritated. The argument seems to be swinging Bohr's way. Now Bohr is the pope of quantum physics, and Einstein is the blasphemer.

Heisenberg will later look back on this conference and construe it as a victory for Bohr, Pauli, and himself—the moment when the 'Copenhagen spirit' descends on quantum physics. The meaning of position and momentum in the atomic world—terms that Einstein himself had redefined only a few years earlier—is now determined by Bohr, Heisenberg, and Pauli. 'I am satisfied in every way with the scientific results,' writes Heisenberg, 'Bohr's and my views are generally accepted. At least, no more serious objections are raised—not even by Schrödinger and Einstein.' He does not mention the fact that his and Bohr's views are not the same. They buried the hatchet in the spring, but their opinions still differ. Heisenberg continues to believe that the world at the atomic scale looks very different from the classical macroworld. And Bohr continues to insist that there is only one world.

The conference comes to a close without a successful rebuttal of quantum mechanics by Einstein. However, he remains steadfast in his opposition to it; anything else would go against the grain for him. Ultimately, it's about his fundamental philosophy. He believes in an objective physical world that develops in space and time according to fixed laws, independently of us humans, but able to be scientifically investigated by us. Quantum mechanics destroys that foundation. For Einstein, it is blasphemy against reality.

When physics enters atomic dimensions, Bohr and Heisenberg claim, the mathematical symbols it uses change their meaning. This is an attack on the physical world as Einstein and others conceived it: he reinvented space and time with his theory of relativity, creating a world of continuity, causality, and objectivity. And now he's expected to sit back and watch as it is torn down again? Never. On the way home from Brussels to Berlin, Einstein takes the train to Paris with Louis de Broglie. 'Don't give up,' he encourages the prince as they take their leave of each other that evening. 'You're on the right track.' But, demoralised by the lack of support he received in Brussels, de Broglie no longer believes that himself.

Einstein returns to Berlin exhausted and downcast, but not converted. A week after the Solvay Conference, on 9 November 1927, he writes to Sommerfeld, 'On quantum mechanics, I think that, with respect to ponderable matter, it contains roughly as much truth as the theory of light without quanta. It may be a correct theory of statistical laws, but an inadequate conception of individual elementary processes.'

The fact that Niels Bohr retains the upper hand in Brussels is due not only to his powers of persuasion, but also to his rallying of his supporters. He learned working in Rutherford's laboratory in Manchester how important a pleasant, communal working atmosphere is, and has always striven since then to maintain the convivial atmosphere among staff at his institute in Copenhagen. Anyone who wants to make it in the quantum science world must go to Copenhagen. 'All roads lead to Blegdamsvej 17,' quips the young Russian quantum physicist George Gamow. By comparison, the Kaiser Wilhelm Institute for Physics in Berlin, run by its founding director, Albert Einstein, is little more than an address. And that's the way Einstein likes it. He prefers to work alone.

Niels Bohr trains an entire generation of theoretical physicists. His students take up the professorships of Europe as they gradually become vacant. Immediately after the Solvay Conference, Werner Heisenberg begins his new job as professor and director of the Institute for Theoretical Physics in Leipzig. Wolfgang Pauli becomes a professor at the Federal Technical University in Zurich, while Pascual Jordan replaces Pauli in Hamburg. Hendrik Kramers took up a professorship in theoretical physics in Utrecht in 1926. These theoreticians exchange students and assistants among themselves. In this way, Bohr's view of quantum physics spreads internationally via his disciples.

One exception to this is Paul Dirac, the mathematical high-flyer among Bohr's students. His memories of the debate between Bohr and Einstein are less rosy. He listened to their arguments, he later reports, but did not take part. 'I was more interested in finding the right equations.' For Dirac, it's clear that quantum mechanics is still an incomplete theory. Einstein may have been put on the defensive in Brussels, but, Dirac believes, he may well turn out to be right.

BERLIN, 1930

Germany flourishes; Einstein falls ill

Berlin, the late 1920s. Germany continues to recover from the ravages of war. The world's largest airport has been built on Tempelhofer Feld, the Berlin parade ground where Orville Wright made his demonstration flight two decades earlier. Fifty planes a day from all over Europe now land at Tempelhof Airport. It also attracts ground-based tourists, who pay a small fee to enter the terminal and can stay as long as they wish. Tourists and locals alike crowd the tables of the terminal cafés, eating, drinking, listening to the hum of the aeroplane engines outside, and watching the shiny planes as they soar high into the sky, or appear out of the clouds, coming from who knows where. Aviation is a great, fascinating adventure. In 1928, aviators John Henry Mears and Charles Collyer make a stop at Tempelhof during their round-the-world flight in their single-engine Fairchild FC-2W. Just outside Berlin, they have to land in a field to ask the local farmers for directions to the city. They spend a couple of hours in Berlin, resting at the Hotel Adlon and breakfasting on eggs, ham, and dark beer, before heading off towards New York, which they will reach again twenty-three days, fifteen hours, twenty-one minutes, and three seconds after setting off. It's a new round-the-world record.

The moribund, nationalist empire has become—at least for a few years—a vibrant, progressive republic. With thirty-six women sitting in the Bundestag, Germany has the highest proportion of female parliamentarians in the world. Women can pursue any profession they choose, at least in principle. They work as engineers, constructors, and butchers. Germany attracts foreigners looking for a classless idyll complete with historic buildings, cobbled streets, beer bars, and relaxed morals. Josephine Baker dances in Berlin, scantily clad in little more than a skirt made of bananas. Max Reinhardt is a star of the stage. The most successful production is Bertolt Brecht's *Threepenny Opera*, with music by Kurt Weill. The Bauhaus movement dominates modern architecture. Scientific endeavour flourishes alongside art and culture.

Various youth movements are spreading through Germany. Bronzed young people populate the pools and beaches, not afraid of showing their bodies. The nature-loving youth movement, the *Wandervögel* (migratory birds), of which Werner Heisenberg is a member, roam across the countryside. On closer inspection, however, the youth movements are less innocent than they first appear. Many young people's organisations are affiliated with political parties.

Albert Einstein, now forty-nine years old, is not doing well. He feels his body is approaching the limits of its capacity. During a short trip to Switzerland in April 1928, he suffers a collapse while lugging his heavy suitcase up a steep hill. Doctors at first fear a heart attack, but then diagnose an abnormal enlargement of the heart. In his words, Einstein feels as if he's 'about to snuff it'. Back in Berlin, Elsa takes charge, limiting access to Einstein and restricting the visits he receives from both friends and colleagues.

While Einstein is gradually recovering, an article by Bohr is published in three languages simultaneously: 'The quantum postulate and the recent development of atomic theory'. In fact, it's a reworking of the lecture he delivered in Como—the last of countless reworkings that Bohr now wants to be seen as the final and eternal version of his interpretation of quantum mechanics, including the concept of complementarity. Bohr sends a copy to Schrödinger, who still refuses to accept the limitations of the uncertainty principle. Replying to Bohr on 5 May 1928, Schrödinger says that if the concepts of position and momentum permit only an imprecise description of a quantum system, then those concepts must simply be replaced by new ones that don't entail such limitations. 'However,' Schrödinger concluded, 'it will no doubt be very difficult to invent this conceptual scheme, since the new-fashioning required touches upon the deepest levels of our experience: space, time and causality.' Thanking him politely in reply, Bohr clarifies that he 'can scarcely be in complete agreement' with Schrödinger's 'emphasis on the need for the development of "new" concepts'. Bohr restates his position that the uncertainty principle is not a more or less arbitrary limitation in the applicability of the classical concepts, but an inescapable feature of complementarity that emerges in an analysis of the concept of observation.

Schrödinger informs Einstein of his exchange with Bohr. And, yes, Einstein agrees, saying that Schrödinger's demand that 'the terms *p* and *q* should be dispensed with' is 'quite justified, if they can describe only such a "wobbly significance"'. Einstein is even willing to throw overboard the traditional basic physical concepts that he himself had redefined, in order not to have to accept the Copenhagen interpretation. 'The reassurance-philosophy (or religion?) of Heisenberg and Bohr has been concocted so subtly that for the time being it provides a soft resting pillow for the believer from which he can no longer be easily roused. So let him lie.'

Four months after his collapse, Einstein is still weak, but no longer confined to his bed. To continue his convalescence, he rents a house in the sleepy village of Scharbeutz on the Baltic coast. There, he spends his time reading Spinoza, his favourite philosopher, and enjoying a break from the 'idiotic existence one leads in the city'. After almost a year, he is finally well enough to return to his office. He works there all morning before going home for lunch and a rest until three o'clock. He then works again, 'sometimes all through the night', recalls his personal secretary, Helen Dukas.

Wolfgang Pauli pays Einstein a visit in Berlin over the 1929 Easter holidays. He's told by Einstein that he still continues to believe in a reality where natural phenomena unfold according to the fixed script of the laws of nature. Pauli finds this attitude 'reactionary'. Einstein's obstinacy in this matter is starting to alienate some of his old friends. 'Many of us consider this a tragedy,' writes Max Born, 'both for him, groping his way forward in solitude, and for us, who miss our leader and champion.'

When he receives the Max Planck Medal of the German Physical Society—awarded to him by Planck himself on 28 June 1928, Albert Einstein makes his views clear in his acceptance speech: 'I admire to the highest degree the achievements of the younger generation of physicists which goes by the name quantum mechanics, but I believe that the restriction to statistical laws will be a passing one.' Einstein has already embarked on the final intellectual quest of his life: his solitary journey in search of a unified field theory that would bring together the theories of electromagnetism and gravity. He hopes in this way to save causality and an observer-independent reality. Who is the reactionary here, Bohr or Einstein? When the two rivals meet again at the sixth Solvay Conference, Einstein will be fully recovered and primed for his next attack on quantum theory.

BRUSSELS, 1930

K.O. in the second round

Brussels, 1930. Monday 20 October sees the start of the sixth Solvay Conference. The six-day event is held under the title 'The magnetic properties of matter'. The procedure is the same as three years earlier. Hendrik Lorentz, who has since died, has been replaced by Paul Langevin as chairman of the organising committee and conference director.

The list of participants is hardly less illustrious than in 1927—with twelve past or future Nobel Prize winners. The participants include Paul Dirac, Werner Heisenberg, and Hendrik Kramers. And, of course, Niels Bohr and Albert Einstein. Arnold Sommerfeld is now also allowed to attend. The ground is prepared for the second round in the fight between Bohr and Einstein over the interpretation of quantum mechanics and the nature of reality itself.

Bohr and Einstein are both well prepared when they arrive in Brussels, like two chess grandmasters before a world championship match. In the intervening three years, Bohr has repeatedly played out in his head the thought experiment Einstein used to try to refute quantum mechanics at the fifth Solvay Conference. He has identified gaps in Einstein's argumentation at the time, but that isn't enough for him. He has dreamt up his own thought experiments, with ever more ingenious arrangements of screens, slits, shutters, and clocks, to test his interpretation of quantum mechanics for weak points. But he has been able to find none.

Once again, Albert Einstein launches an all-out attack on quantum mechanics, and Niels Bohr once again defends the theory. Bohr considers himself prepared for anything, until, after one of the official meetings, Einstein conjures up a light-filled box. An imaginary box. Einstein has come up with a devilish thought experiment. Imagine a box, Einstein tells Bohr, that contains light particles and a clock. In one of the walls of the box is a hole with a shutter that can be opened and closed by a mechanism connected to the clock inside the box. Weigh the box. Set the clock to open the shutter for the briefest of moments, just long enough for only a single photon to escape. We now know, explains Einstein, the precise time when the photon left the box. Bohr listens, as yet unconcerned. Everything so far seems straightforward and beyond contention. The uncertainty principle applies only to pairs of complementary variables, for example position and momentum, or energy and time. It does not impose any limit on the degree of accuracy with which either one of the pair can be measured. Then Einstein plays his trump card. 'Weigh the box again.' In a flash, Bohr realises that he's now in deep trouble.

Einstein is using the great discovery he made while still a clerk at the Patent Office in Bern: energy is mass, and mass is energy. $E=mc^2$. The difference in weight of the box before and after the shutter has been opened and closed is a measure of the energy of the escaped particle. Such a small change would be impossible to measure using equipment available in 1930, but that's a technical irrelevance, as this is a thought experiment dealing with principles, not practicalities. The time of the particle's escape is known, and so is its energy. Both at the same time. Which contradicts the uncertainty principle. Can quantum mechanics be outwitted after all?

Bohr is taken aback. He sees no way out of this; he has no answer. Pauli and Heisenberg try unsuccessfully to reassure the doyen of quantum mechanics, telling him, 'It can't be right, everything will be fine.' It's a nice gesture, but not very helpful. Bohr spends the whole evening in discussion with his followers, saying it simply cannot be true, and warning them that this could mean the end of physics. Or, at least, the end of physics as Bohr understands it.

That evening, a silent Einstein strides back to the Hotel Metropole, tall and majestic, puffing away on a cigar, with a subtle smile on his lips. Beside him scurries a very agitated Bohr, arguing, gesticulating, sweating, with his coat over his arm, and looking 'like a dog who has received a thrashing', in the words of his Belgian friend and colleague Léon Rosenfeld.

For Einstein, it doesn't mean the end of physics at all, but rather its salvation and the rescue of an observer-independent reality: physics as Albert Einstein understands it.

As others slumber, Bohr spends a sleepless night examining every facet of Einstein's light box. He mentally takes the imaginary box of light apart in search of a flaw, visualising it in far more detail than Einstein himself ever did. In his mind, Bohr suspends the light box from a spring fixed to a supporting frame with a scale that shows its vertical position. He starts the clock mechanism ticking, and mentally designs the mechanism that opens and closes the escape hole. He even considers the nuts and bolts used to fix the frame to the base. Sometimes a mind's greatness is shown in its ability to focus on the smallest of details.

The next morning, an exhausted but confident Niels Bohr enters the breakfast room at the Metropole—no longer anyone's idea of a freshly thrashed dog. He immediately launches into an exposition of the results of his night-time contemplations. Einstein has forgotten something: during the weighing process, the box moves a tiny amount in the gravitational field of the Earth. That causes a small amount of uncertainty in its measured mass, and therefore in the energy of the escaped particle. In addition, the speed at which a clock runs differs depending on its position in the Earth's gravitational field. Einstein himself had shown this years before. So the time measurement is also inaccurate. Taken together, these two inaccuracies constitute precisely the uncertainty that the principle predicts.

The morning coffees are left to go cold on the breakfast table. What a turnaround! Bohr has transformed Einstein's supposed rebuttal into a brilliant confirmation of the uncertainty principle using an ingenious application of Einstein's own theory of relativity.

Now it's Einstein who is stunned into silence, stumped for an answer. Just as he did three years earlier, Bohr has demolished Einstein's line of attack. Einstein could still ask how quantum mechanics can be a self-contained theory if it needs to resort to the theory of relativity for salvation, but he offers no more arguments. Now is the time to accept defeat. For the time being, at least. This is to be the last public debate between Bohr and Einstein. But it will not be Einstein's final attack on quantum mechanics. He changes tack and no longer concentrates on trying to circumvent the uncertainty principle. Instead, he focuses on exposing another point of weakness, which he calls 'spooky action at a distance'.

In November 1930, Albert Einstein gives a lecture on his light-box experiment at the University of Leiden, where he often appears as a guest speaker. After his lecture, one of the audience members comments that he sees no conflict with quantum mechanics. 'I know,' replies Einstein. There are no contradictions there. But the theory is wrong nonetheless, Einstein insists.

Einstein is stubborn, but not petty. Despite his aversion to quantum mechanics, he nominates Werner Heisenberg and Erwin Schrödinger once again for the Nobel Prize in 1931. In his nomination letter, he writes, 'In my opinion, this theory contains without doubt a piece of the ultimate truth.' A piece of the truth. Not the whole truth. Einstein's 'inner voice' continues to whisper to him that quantum mechanics is not the answer to everything, as Bohr believes.

After the sixth Solvay Conference, Einstein spends a couple of days in London. On 28 October, he is the guest of honour at a charity dinner at the Savoy Hotel, in aid of poor Eastern European Jews, which is organised by the Joint British Committee of ORT-OZE Societies. The dinner is hosted by the committee's president, Baron Rothschild, and the master of ceremonies is George Bernard Shaw. There are almost a thousand guests in attendance. Einstein feels ill at ease among the rich and beautiful, dressed in their finery and dripping with jewellery and medals. But he's prepared to endure the 'monkey comedy' for a good cause, squeezing himself into a tailcoat and white tie, and shaking hands with those queueing to meet the 'Jewish saint'.

The seventy-four-year-old Shaw rises and proposes a toast to 'Professor Einstein's health'. There are men such as Napoleon, who are makers of empires, Shaw says, but there are men who go far beyond that. They are makers of universes. And their hands are not stained with the blood of their fellow men. 'I can count them on the fingers of my two hands,' Shaw continues, 'Pythagoras, Ptolemy, Kepler, Copernicus, Aristotle, Galileo, Newton, and Einstein ...' he recounts, to thundering applause from the guests, 'and I still have two fingers left vacant,' he continues, to laughter from the audience, while a bittersweet smile crosses Einstein's face. 'Ptolemy made a universe, which lasted 1,400 years. Newton, also, made a universe, which lasted for 300 years. Einstein has made a universe, and I can't tell you how long that will last.' Few of those present are aware that Einstein is currently fighting to save the foundations of the universe he's made.

After Shaw's introduction, Einstein rises to speak, thanks Shaw for 'the unforgettable words which you have addressed to my mythical namesake who makes life so difficult for me'—and makes no further mention of science. He concludes his speech with the words, 'To you all I say that the existence and destiny of our people depends less on external factors than on our remaining faithful to the moral traditions which have enabled us to survive for thousands of years despite the fierce storms that have broken over our heads.' Six weeks earlier, on 14 September 1930, 6.4 million Germans had voted for the NSDAP, the Nazi Party, in the Reichstag elections, which was eight times more than in the previous election of May 1928, to the dismay of many moderate citizens. With this result, the Nazis proved that they were more than just one more far-right fringe group. They now make up the second-largest block in the German parliament, with 107 seats. The Social Democrat–led grand coalition is fractured, and chancellor Heinrich Brüning can only govern as the head of a minority government and by emergency decree. On 13 October, the NSDAP parliamentarians turn up to the first session of the new Reichstag wearing their brown party uniforms. On the same day, Jews are chased through the streets of Berlin, abused, and beaten. The windows of the Jewish-owned Wertheim department store are smashed. Rumours of an imminent right-wing coup begin to circulate. The Weimar Republic is reeling.

Albert Einstein sees all this—the votes for Hitler and the anti-Semitic attacks—as symptoms of the deep and spreading fear, desperation, and insecurity caused by economic misery and unemployment. Between the 1928 elections and those in 1930, the world has suffered the great Wall Street Crash.

The crash hits the whole of Europe, but Germany suffers more than most. Much of the country's economic recovery since the First World War has been funded by borrowing—often with short-term loans from the United States. With increasing chaos and mounting losses, American banks demand the immediate repayment of those loans. Almost no foreign capital now flows into Germany. The result is a rapid rise in unemployment, from 1.3 million in September 1929 to more than three million in October 1930.

Einstein sees the rise of the Nazis and the persecution of Jews as a 'temporary result of the current desperate economic situation' and a 'childish disease of the Republic'. Temporary. Then he's forced to watch as the childish disease becomes life-threatening and eventually kills off the republic, rather than strengthening its immune system. Germany's first republic is now nothing more than a shell of words. 'Political authority emanates from the people' is the first article of the Weimar Constitution, but the republic is now in effect being governed by decree. The democratically elected parliament is powerless.

Politically aware people are growing increasingly concerned. 'We are moving toward bad times,' writes Sigmund Freud to Arnold Zweig in December 1930. 'I ought to ignore it with the apathy of old age, but I can't help feeling sorry for my seven grandchildren.' In his diary in the following years, Freud records the ever-increasing signs of a move to the right and attacks on Jewish citizens.

Many people are unsure what to think of these developments. They may not like the Nazis, but wouldn't the Bolsheviks be even worse?

Germany's physicists react differently to the rise of Nazism. Some passionately oppose it. Some flee the country. Some turn a blind eye, or try to find a way to live with it. Some are happy to join in. Philipp Lenard and Johannes Stark, both Nobel laureates, reveal themselves to be anti-Semites. They see an opportunity to develop a 'German physics', opposed to both the theory of relativity and quantum mechanics, which they consider too obscure, too mathematical, and generally too 'Jewish'.

Albert Einstein reluctantly has to accept that there is no place for him in a country run by Nazis. At the beginning of December 1930, he leaves Germany to spend two months at Caltech in Southern California, which has developed over the past few years into one of the most important centres of scientific research in the US. Ludwig Boltzmann, Hendrik Lorentz, and Erwin Schrödinger have all delivered lectures at Caltech. When Einstein's ship docks in New York, he is persuaded to hold a fifteen-minute press conference for the hordes of reporters waiting there. 'What do you think of Adolf Hitler?' shouts one. 'He is living on the empty stomach of Germany,' replies Einstein. 'As soon as the economic conditions improve, he will no longer be important.'

In December 1931, Einstein leaves for another stint at Caltech. The economic and political situation in Germany has deteriorated further. As he crosses the Atlantic, Einstein writes in his diary, 'I decided today that I shall essentially give up my Berlin position and shall be a migratory bird for the rest of my life!'

While in California, Einstein happens to meet the educational reformer Abraham Flexner, himself the son of German immigrants. Flexner is in the process of establishing a unique research centre, the Institute for Advanced Study, in Princeton, New Jersey. Armed with a $5 million donation, Flexner wants to create a 'society of scholars' devoted entirely to research, free of the demands of teaching students. Having met the world's most celebrated scientist by chance, Flexner is not about to waste the opportunity to recruit Einstein. In 1933, Einstein becomes the first professor at the Institute for Advanced Study—a position he will retain for the rest of his life.

Einstein negotiates an arrangement under which he will spend five months a year in Princeton and the rest of the time in Berlin. 'I am not abandoning Germany,' he stresses to the *New York Times*. 'My permanent home will still be in Berlin.' The agreement is initially for five years and begins in 1933, after a third stint at Caltech to which Einstein has already committed himself. And it's lucky he has, because during his time in Pasadena, on 30 January 1933, Hitler will be appointed Reich chancellor.

On 10 December 1932, the Einsteins board the SS *Oakland* in Bremerhaven, along with thirty pieces of luggage.

In the safety of California, Einstein does not speak out, but he acts as if he will return when the time comes, even writing to the Prussian Academy of Sciences to inquire about his salary. But he has already secretly made his decision. 'In view of Hitler,' he writes to Margarete Lebach in Berlin on 27 February, 'I don't dare step on German soil. I have already cancelled my lecture at the Prussian Academy of Sciences.' That same night, the Reichstag building goes up in flames, heralding the start of the first wave of Nazi terror against left-wing politicians, intellectuals, and journalists.

On 10 March 1933, the day before his planned departure from Pasadena, Albert Einstein releases a statement and gives an interview explaining his view of the events in Germany: 'As long as I have any choice in the matter, I shall live only in a country where civil liberty, tolerance and equality of all citizens before the law prevail. Civil liberty implies freedom to express one's political convictions; tolerance implies respect for the convictions of others whatever they may be. These conditions do not exist in Germany at the present time. There, those are persecuted who have contributed particularly to fostering international understanding, including some leading artists.' The journalist to whom Einstein gives the interview sees the ground shaking beneath his feet as he walks across the Caltech campus. On 11 March, Einstein and his wife leave Pasadena, uncertain where their next journey will take them.

Einstein's words cause a sensation around the world. In many places, there is ambiguity over Hitler. Some people outside Germany, while acknowledging his ruthless treatment of the Jews, weigh that against his attempts to restore national confidence among Germans, and consider him a protector of free Europe against the Bolshevik threat. Is it such a high price to pay, as Einstein claims? German newspapers seize the opportunity to demonstrate their allegiance to the Führer, with ostentatious declarations of outrage at Einstein's statement. 'Good News from Einstein—He's Not Coming Back!' is the headline in the *Berliner Lokal-Anzeiger*, and the *Völkischer Beobachter* even publishes pamphlets aimed at stirring up hatred against him.

Einstein's statement against the Nazis leaves Planck in a quandary, as he attempts to navigate through the political turmoil. On 19 March 1933, Planck writes to Einstein of his 'profound distress' over 'all kinds of rumours that have emerged in this unquiet and difficult time about your public and private statements of a political nature. I am not in a position to verify their significance. The only thing I see very clearly is that these reports make it exceedingly difficult for all those who esteem and revere you to stand up for you.' It's another iteration of the often-used argument to discourage people from taking a stand against the Nazis: Think of your loved ones; don't put them in danger! It's a perverse reversal of cause and effect, in which Planck blames Einstein for making the difficult situation of his 'tribal companions and co-religionists' worse.

On 28 March 1933, while on board the steamship *Belgenland*, Einstein writes a letter to his son Eduard in Zurich: 'For the time being, I will not be returning to Germany, perhaps never again.' He pens another letter, this one to the Prussian Academy of Sciences, resigning his position 'due to the conditions currently prevailing in Germany'—much to Planck's relief, who had feared being forced out of his cowardly position by taking part in an official expulsion procedure against Einstein, which would have made him take an unambiguous stance on the matter.

Immediately after disembarking in Antwerp, Einstein asks to be driven to the German embassy in Brussels. There he surrenders his passport and renounces his German citizenship. He spends the summer of 1933 in Belgium and Oxford.

The Nazi Party declares a 'Jew boycott' for 1 April of that year. SA Stormtroopers are posted outside Jewish-owned shops, and Jewish students, assistants, and lecturers are prevented from entering universities and robbed of their library cards. The perpetual secretary of the Prussian Academy of Sciences, Ernst Heymann, issues a declaration on behalf of the academy, accusing Einstein of 'atrocity mongering' and stating that the academy 'therefore has no reason to lament Einstein's departure'. The only academy member to contradict this sentiment is Einstein's old friend Max von Laue. The academy does not have to be brought into line by the Nazis—it toes the line voluntarily.

In a letter to Planck from the Belgian coast dated 6 April 1933, Einstein gives his opinion of his former colleagues' behaviour: 'I will give the academy the benefit of assuming it made these slanderous statements only under outside pressure. But even if that should be so, its conduct will hardly be to its credit; some of its more decent members will certainly feel a sense of shame …'

Albert Einstein doesn't know where he should settle. He receives job offers from universities around the world, but which should he accept? He and Elsa move into a villa in Coq-sur-Mer on the Belgian coast for six months. Rumours of a German plot to murder him reach Belgium. The *New York Times* reports that the SA has searched Einstein's house in Potsdam and confiscated a breadknife. The Belgian government assigns two guards to protect him.

In September 1933, fears for Einstein's safety in Belgium reach such a level that he moves to England, spending a peaceful month in a small cottage on the Norfolk coast.

Einstein is persuaded to make a speech at a charity event in aid of refugees in need. Ernest Rutherford hosts the event, which takes place at the Royal Albert Hall on 3 October 1933. Einstein delivers the speech in his Swabian-accented English, hesitantly reading aloud from a piece of paper, but that does nothing to dampen the enthusiasm of the audience. Ten thousand people throng the hall, anxious to see and cheer him. At the request of the organisers, Einstein never once mentions the word 'Germany', although his entire speech is about the situation there and the danger it spells for the rest of the world.

Four days later, on 7 October 1933, Einstein leaves for America. He boards the *Westmoreland* in Southampton, joining his wife Elsa and his secretary Helen Dukas, who had embarked in Antwerp. The plan is to spend only the next five months at the Institute for Advanced

Study; in fact, Einstein will never return to Europe again.

While still in quarantine, Einstein receives a letter from Abraham Flexner, the founder and director of the institute, entreating him to 'remain silent, exercise discretion, and refuse public appearances' for his own safety. 'No doubt there are organised bands of irresponsible Nazis in this country,' writes Flexner. It doesn't take a brain the size of Einstein's to realise that Flexner's real concern is the damage that Einstein's public statements might inflict on the reputation of his fledgling institute, and on the donors it relies upon. Over the next few weeks, Flexner habitually intercepts Einstein's mail and turns down invitations and appointments in his name—including an invitation to visit the White House. Einstein becomes increasingly frustrated at being muzzled in this way by Flexner. It was precisely that kind of treatment he had left Germany to avoid. In letters to friends, he even gives his return address as 'Concentration camp, Princeton'.

Einstein writes to the trustees of the institute to complain about Flexner's behaviour, listing his misdeeds and asking them to guarantee him 'security for undisturbed and dignified work, in such a way that there is no interference at every step of a kind that no self-respecting person can tolerate. If this is not considered possible, then I will have to discuss with you ways and means of severing my relations with your institute in a dignified manner.' The threat works. A dissatisfied Einstein would be even more damaging to the institute than an outspoken Einstein. Flexner has to leave him alone. Einstein gains the right to do as he pleases, but at a price. He will never have any real influence in the running of the institute. It is the freedom of a king's fool.

ZURICH, 1931

Pauli's dreams

Zurich, summer 1931. Passing through a garden gate, Wolfgang Pauli approaches a grand house by the lake. There is a strange contrast between the outside world and his inner life. The villa, with its turrets, gables, and landscape windows, its fruit trees and neatly trimmed bushes, and the sailing boats on the lake all make for an idyllic, picture-book scene. But inside Wolfgang Pauli, a dark storm is raging. Pauli, the highly gifted physicist, is struggling to cope with life. That's what has brought him here, with anxious hope, to this lakeside villa. It's the home of Dr Carl Gustav Jung, the great psychiatrist and a one-time student of Sigmund Freud, but now his greatest adversary. This is Pauli's first appointment with Jung.

Wolfgang Pauli is one of the greatest talents in the history of physics. His teacher Max Born describes him as 'a genius, comparable only to Einstein himself. As a scientist he was, perhaps, even greater than Einstein.' By which, Born means Pauli is anything but a genius when it comes to people. Often in conflict with his colleagues, Pauli has a bitter sense of humour that often offends people. He has no luck with women, and women are lucky to avoid him. And he has an alcohol problem.

Despite all this, Pauli is valued by his colleagues. His criticism is usually on point, and he wastes no time getting to that point. 'Perhaps more important than his publications are the innumerable, untraceable contributions, advancing the development of the new physics by oral contributions to the discussion or by letters,' says his friend Paul Ehrenfest. But not even Ehrenfest, not even Pauli's closest friends, realise the tragedy that is playing out behind the scenes.

Wolfgang Pauli was born in Vienna on 25 April 1900, at a time when Vienna's vibrancy was tinged with *fin-de-siècle* angst. Pauli's father, also called Wolfgang, was a doctor before changing from medicine to science. In the process, he also changed the family name—from Pascheles to Pauli—and converted from Judaism to Catholicism, fearing that the growing atmosphere of anti-Semitism could damage his academic career.

Wolfgang Junior grows up knowing nothing of the family's Jewish background. It's not until a taunt from a schoolmate prompts him to question his parents that he finds out the truth. Pauli's father sees his decision to assimilate the family as vindicated when he is awarded the professorship for biological and physical chemistry, as well as the directorship of the corresponding institute at the University of Vienna. After the German 'annexation' of Austria in 1938, Pauli is classified as a Jew under the Nazi race laws, and is forced to flee the country. He moves to Zurich.

Pauli seems to have been destined for life in physics from an early age. His godfather is Ernst Mach, the leading Viennese physicist and philosopher. Pauli later describes his relationship with Mach, whom he last sees when he's fourteen years old, as 'the most important event in my intellectual life'. Pauli is soon recognised as a child genius, about which he will later comment, 'Ah, the child genius—the genius passes and the child remains ...'

Wolfgang Pauli's mother, Bertha, is a prominent journalist in Vienna, known as a pacifist, a socialist, and a women's rights activist. The Paulis' house is a meeting place for artists, scientists, and medics. Bertha has a strong influence on Wolfgang's opinions, especially during the First World War. The longer it drags on, the greater the teenage Wolfgang's opposition to it becomes.

Pauli is highly gifted, but he isn't a model pupil. School bores him. During particularly dull lessons, he secretly reads Albert Einstein's works on the theory of relativity, hiding the books under his desk. There is a story that once, during a physics lesson at school, the teacher made a mistake in his calculations on the board, but was unable to find the error, no matter how much he recalculated. To the great amusement of Pauli's classmates, the exasperated teacher finally called out, 'Come on, Pauli, tell me where the mistake is. I know you spotted it a long time ago.'

At the age of eighteen, Pauli flees the 'spiritual wasteland' of Vienna, as he calls it. The Austro-Hungarian Empire is heading towards its demise, the glory of its capital is fading, and all the good physicists are leaving the university. Pauli goes to Munich to study under Arnold Sommerfeld, who has just refused a professorship in Vienna. Sommerfeld is in the process of turning Munich into a 'seedbed of theoretical physics' where the best teachers train the greatest talents. He has been trying to achieve this since 1906, but his institute is still small and familial. It consists of just four rooms: Sommerfeld's office, a lecture theatre, a seminar room, and a small library. There is a laboratory in the basement, which is where Max von Laue's theory that X-rays are high-energy electromagnetic waves was confirmed in 1912.

Sommerfeld is an excellent theoretician and an even better teacher. He has a natural ability to set problems for his students that will stretch their abilities, but not frustrate them. Although he has already taught many talented students, he quickly realises that Pauli is a student of extraordinary ability.

Pauli moves to Munich straight after completing his school leaving exams in 1918, to be cultivated in Sommerfeld's 'seedbed'. During Pauli's third semester, Sommerfeld asks him to write the chapter on the theory of relativity in the *Encyclopaedia of Mathematical Sciences*. Einstein himself doesn't want to write it, Sommerfeld doesn't have the time, and Pauli is, of course, already very familiar with the theory.

Just under a year later, the chapter is finished: it is 237 pages long, with 394 footnotes. Pauli writes it alongside his studies. Einstein is delighted. 'Whoever studies this mature and grandly composed work would not believe that the author is a man of twenty-one. One does not know what to admire most: the psychological understanding of the evolution of ideas, the accuracy of mathematical deduction, the deep physical insight, the capacity for lucid, systematic presentation, the knowledge of literature, the factual completeness, or the infallibility of criticism.' The chapter remains a standard reference work on the subject of relativity for decades.

Wolfgang Pauli is soon known and feared for his sharp, incisive, and uncompromising criticism of new, speculative ideas. His good friend Paul Ehrenfest nicknames him 'the Scourge of God', while others call him 'the conscience of physics'. He dismisses the work of one young physicist as 'not even wrong'. When a colleague tries to slow him down by saying, 'I can't think as fast as you,' Pauli answers, 'I don't mind if you think slowly, but I do have something against it when you publish faster than you think.' Remarks like this earn Pauli a reputation for arrogance. Those who know him better also know that he's a straight-talking person, but not a hurtful one. 'Pauli was an exceedingly honest person,' says his colleague Victor Weisskopf, 'he had a kind of child-like honesty. He always voiced his genuine opinion straight out, without any inhibition.'

Pauli's fellow physicists like to tell a joke about him: 'After Pauli dies, God grants him an audience. Pauli asks God why the fine-structure constant has a value of 1/137. Nodding, God goes to the chalkboard and starts furiously deriving equation after equation. At first, Pauli looks on with great pleasure, but then begins to shake his head vehemently …'

Pauli has a habit of rocking his body back and forth when he's concentrating. His intuition for physics is unrivalled among his contemporaries. Even Albert Einstein cannot surpass it.

Pauli criticises his own work even more harshly than that of others. Sometimes it seems that his knowledge of physics and its problems is so great that it hinders his creativity. He lacks the carefree naivety that great intellectual explorers sometimes need. So he fails to make the discoveries that his intuition and imagination should lead him to, and they are eventually ascribed to less talented, but also less cautious, scientists.

Even when it comes to Einstein, Pauli remains confident and unabashed. While still a student, Pauli tells a packed lecture hall, 'What Einstein says is not completely stupid,' while Einstein—who has just given a guest lecture in Munich—is also in the audience.

Pauli spares no one. Except one person. The only individual with whom Pauli bites his razor-sharp tongue is Sommerfeld, whom he always addresses as 'Mr Privy Councillor'. When Sommerfeld speaks, Pauli listens quietly and humbly. 'Yes, Mr Privy Councillor; no, Mr Privy Councillor, perhaps it's not quite as you say.'

As a student, Pauli makes the most of Munich's nightlife. In the evenings, he hangs around in cafés till closing time, before going back to his room to spend the rest of the night working. He usually doesn't bother attending morning lectures, turning up at around midday. But he hears enough to be drawn into the mysteries of quantum physics by Sommerfeld. 'I was not spared the shock,' he later says, 'which every physicist accustomed to the classical way of thinking experienced when he came to know of Bohr's basic postulate of quantum theory for the first time.'

For Pauli's doctoral thesis, Sommerfeld sets him the task of applying the principles of the Bohr-Sommerfeld atomic model to hydrogen molecules that have been ionised by the removal of an electron from one of their two atoms. Pauli delivers a theoretically perfect analysis, which, however, does not match measurements made in the lab. This is not his fault, as it derives from the fact that the limits of the predictive power of Bohr and Sommerfeld's theory have been reached. Pauli is awarded his doctoral title. In October 1921, the year of his doctorate and the publication of his article on the theory of relativity, Pauli heads to Göttingen to take up a position as Max Born's assistant. The thirty-eight-year-old Born has been teaching in the small university town for only six months.

Born is very impressed by 'little Pauli'. 'I will never find a better assistant,' he writes to Einstein, shortly after Pauli's arrival. Together, Born and Pauli work meticulously to find a way to apply the methods of celestial mechanics to atoms and molecules. One thing they have in common is their terrible clumsiness. In fact, Pauli is even more inept in the laboratory than Born.

Bad luck seems to follow Pauli around. His colleagues, especially the experimental physicists among them, speak of the 'Pauli effect'. Physicists have a theory about their kind, which says that there is a 'law of the conservation of genius' separating theoreticians and experimentalists. All brilliant theoreticians are terrible at experiments, and vice versa. Wolfgang Pauli is living proof of this theory. His genius is purely theoretical. A superstition arises to the effect that, wherever Pauli appears, something is bound to break. If he visits an observatory, its large refracting telescope will suddenly be badly damaged. One time, in the laboratory at Göttingen, a whole complex experimental apparatus for studying atoms collapses for no discernible reason. How can this have happened, the experimental physicists ask themselves, when Pauli is far away in Switzerland? The laboratory manager sends a humorous letter about the incident to Pauli's address in Zurich. The reply arrives with a Danish postmark. Pauli is in Copenhagen. The very moment the measuring instruments failed was precisely the time that Pauli's train to Denmark pulled into the station at Göttingen. In Hamburg, the city's leading experimental physicist refuses to let Pauli into his lab, communicating with him only through the closed laboratory door—out of concern for the safety of his equipment.

Born has to accept his highly gifted assistant's idiosyncrasies. Pauli's phenomenal intellect is still mainly active at night. He is still in the habit of working late and getting up late in the morning. When Pauli is due to stand in for his boss at an eleven o'clock lecture, Born has to send his housemaid to wake Pauli at half past ten.

Anyway, Pauli is Born's 'assistant' in name only. Despite the erratic lifestyle and chronic lack of punctuality of the 'child genius', Born has to admit that he is learning far more from Pauli than Pauli could ever learn from him. Pauli has an unfailing physical intuition for pursuing the right course, which Born can only achieve through hard mathematical work. The teacher is more regretful than the student when their ways part after only one semester. Pauli moves to Hamburg in April 1922, taking him from a small university town to a major city, 'from mineral water to champagne', as Pauli himself puts it. He spends his nights enjoying the pleasures of St. Pauli, the famous drinking and red-light district, with his new clique of friends: the physicist Otto Stern, the mathematician Erich Hecke, and the astronomer Walter Baade. It's an alcohol problem waiting to happen.

Two months later, Pauli returns to Göttingen to attend the 'Bohr Festival', the prelude to the most exciting period for quantum mechanics. This is where he and Bohr meet for the first time. Pauli also makes the acquaintance of Werner Heisenberg, the other rising star of quantum physics. The great Bohr asks Pauli if he wants to come to Copenhagen. Of course Pauli does, and he expresses his enthusiasm in his inimical way, saying to Bohr, 'I hardly think that the scientific demands you will make on me will cause me any difficulty, but the learning of a foreign language like Danish far exceeds my abilities.' Pauli arrives in Copenhagen in the autumn of 1922. Both his assumptions turn out to be wrong. The language doesn't pose much difficulty for him. But the science does.

In Copenhagen, Pauli sets about analysing the 'anomalous Zeeman effect', a phenomenon affecting atomic spectra that cannot be explained by the Bohr-Sommerfeld model of the atom. When atoms are exposed to a magnetic field, their spectral lines split. One line becomes two, three, or even more. Arnold Sommerfeld has cleverly modified Bohr's model to explain the doubling and tripling of spectral lines. The magnetic field stretches, distorts, and twists the electrons' orbits. Sommerfeld describes these effects with three 'quantum numbers'. But the lines can split even further, into four, or six, which is a phenomenon that Bohr's mix of classical physics and quantum theory cannot explain.

Pauli aims to rectify this situation, but fails. Again and again, he finds himself lost in a labyrinth of false paths. 'It simply refuses to be right!' he complains to Sommerfeld. 'So far I've got it totally wrong!' He gradually descends into an ever-deepening state of despair. His time in Copenhagen is running out. As he wanders aimlessly through the streets of the city, a colleague encounters him and asks him why he's looking so glum. 'How can one look happy when thinking about the anomalous Zeeman effect?' is his stern reply. He suspects that 'in the anomalous Zeeman effect there is no conditionally periodic model and something essentially new must be done'.

Pauli returns to Hamburg without having solved the Zeeman-effect problem. On the positive side, he's now been promoted from assistant to university lecturer. He can't stop thinking about the riddle of the Zeeman effect, and he can't stop thinking about Copenhagen. He often takes the train and the ferry across the Baltic Sea. He continues to search for a solution. What is missing from Bohr's atomic model? As his desperation grows, so does his alcohol problem. Problem? Yes, he likes a drink, but it's not a problem ... Pauli remarks that 'drinking wine agrees very well with me. After the second bottle of wine or champagne I usually adopt the manners of a good companion (which I never have in the sober state) and then may at times enormously impress those around me, especially if they are female!' It's the start of a double life: by day, Pauli is a conservative university lecturer; by night, he's a man about town. After dark, he roams the bars and music halls of St. Pauli, where the counters are sticky with spilled beer, and the walls are yellow with tobacco smoke, and where Josephine Baker dances her Charleston, now she's been banned from performing in Munich. Pauli keeps his nightly forays into the underworld of St. Pauli secret from his colleagues.

Pauli's impression that alcohol improves his manners is wrong. Alcohol lowers his inhibitions, and he often becomes violent, even getting involved in fights. Once, while eating in one of his favourite haunts, he finds himself right in the middle of a fight. Flying into a rage, he only comes to his senses when someone threatens to throw him out of a second-floor window. Later, he can't understand his own behaviour. He has affairs with women, realising too late that they are drug addicts who are only with him to get money for their fix. Pauli is gradually losing control.

In the autumn of 1924, there are also developments in Pauli's diurnal life. A tip from Arnold Sommerfeld puts him on the right track. In the fourth edition of his textbook, *Atomic Structure and Spectral Lines*, Sommerfeld mentions a paper by the thirty-five-year-old English long-term student Edmund Stoner, published in the *Philosophical Magazine*. Stoner, who works as an experimental physicist at Rutherford's Cavendish Laboratory, claims that in large atoms, electrons are arranged in a different configuration from that described by Bohr. Bohr contests Stoner's claim. Pauli is immediately enthused.

Though not known for his sporting prowess, he leaps up from reading Sommerfeld's book and sprints to the library in search of a copy of the *Philosophical Magazine*. Now he understands what happens in the shells of atoms. The answer lies in a fourth quantum number. Electrons have a property that no one has yet recognised—a property that Pauli calls 'ambiguous'. It can only have one of two values: 0 or 1. This doubles the number of possible states for an electron in an atom—and suddenly, everything fits together. The distribution of electrons is now as it should be. And the spectral lines should split just as experiments have shown they do for decades.

The key to Wolfgang Pauli's explanation of the arrangement of electrons in atoms is the principle that will forever bear his name: the Pauli exclusion principle. It states that two electrons cannot occupy the same state simultaneously in an atom. It is one of the great laws of nature. It enables Pauli to explain why the elements in the periodic table are arranged in the order they are, and how electrons are configured in the atoms of noble gases. He can now explain the structure of matter itself. But he's not able to explain his own principle. It's a fundamental, underlying principle of nature.

Pauli has finally achieved his goal, even if that goal found him, rather than him finding it. In his paper 'On the connection between the completion of electron groups in an atom with the complex structure of spectra', published in the *Zeitschrift für Physik*, he admits, 'We cannot give a further justification for this rule, but it seems to be a very natural one.'

Pauli has now shown what he's capable of, and continues to do so. He's involved in the development of Heisenberg's matrix mechanics, proves the equivalence of matrix and wave mechanics, and helps Heisenberg formulate his uncertainty principle. Through all of that, he remains the rebellious, headstrong person he's always been. He refuses to submit to either the 'numerical mysticism of Munich' (Sommerfeld) or the 'reactionary Copenhagen coup' (Bohr).

In 1930, Wolfgang Pauli boldly predicts the existence of a mysterious, as-yet-unidentified particle—with no charge, almost no mass, and an ability to pass through almost any measuring device undetected. He calls it a 'neutron', later renamed the 'neutrino'. It's a tremendous advance, as, until now, physicists have considered matter to be made up exclusively of electrons and protons. Twenty-six years later, the existence of the neutrino is proven.

Pauli is at the high point of his scientific career, but his personal life is at a low point. His mother had killed herself by poisoning in 1927, after discovering Pauli's father was having a sexual affair. His emotions in turmoil, Pauli married a cabaret dancer from Berlin by the name of Käthe Deppner in 1929. Two months after their wedding, Pauli wrote in the postscript to a letter to a friend, 'In case my wife runs away, you (just like all my other friends) will receive a printed notice.' Only a few months later, they divorced after Käthe ran away with a chemist. 'If she had taken a bullfighter, I would have understood, but an ordinary chemist …'

Now Pauli drowns his disappointment in drink, slipping ever deeper into crisis. He puts on weight, and his face becomes swollen. During a lecture tour of the US, he manages, despite prohibition, to get so drunk that he falls down a flight of stairs and breaks his shoulder. With a splint on his right arm, he's dependent on an assistant to write on the board for him.

Back in Zurich, where he has been based since accepting the professorship in theoretical physics at the Federal Technical University in 1928, Pauli continues to drink, and also takes up smoking. He staggers from one party to another, having random sex with women, and getting into fights with men. The administration at the institute calls him in for an interview. Pauli is warned that his job is in danger unless his behaviour improves. What use is being 'an even greater genius than Einstein' to a person who is depressed? The mask of the jolly, witty professor hides emptiness and desperation. The disturbing dreams that have long plagued Pauli at night now also haunt him during the day. They involve geometric shapes, combined with images from the world of physics such as clocks and pendulums, and mysterious symbols such as the serpent that swallows its own tail, or chimaeras made up of human and animal bodies, and naked, half-veiled female figures. He can no longer drink the images away.

Wolfgang Pauli has hit rock bottom.

At his father's suggestion, Pauli contacts the world-famous analyst Carl Jung, who lives in Zurich, almost in Pauli's neighbourhood. Although Pauli feels more hate than love for his father, he follows his advice and makes an appointment to see Jung.

True to his personality, Pauli turns up for his appointment well prepared, having read many of Jung's writings beforehand. Pauli underlines one passage in Jung's book *Psychological Types* three times: 'If the persona is intellectual, the anima will certainly be sentimental. A very feminine woman has a masculine soul, and a very manly man a feminine soul. This opposition is based upon the fact that a man, for instance, is not in all things wholly masculine, but has also certain feminine traits.' Pauli is an intellectual, there's no doubt about that. Do his problems stem from his hidden, feminine side?

An unfamiliar feeling rises in him. It's the feeling that the inner contradictions pulling him in different directions have finally been understood. But he also feels a sense of déjà vu. Jung's theory of the 'union of opposites', his attempt to unify the un-unifiable within human beings, reminds Pauli of quantum physicists' struggle to understand the nature of matter. It puts him in mind of Bohr's theory of complementarity. Is the relationship between the feminine and the masculine comparable to waves and particles?

For Pauli, who is so skilful in covering up his insecurity with wit and ridicule, reading Jung's description of the introverted thinking type is like looking at himself in a mirror: 'His judgment appears cold, inflexible, arbitrary, and ruthless; however clear to him the inner structure of his thoughts may be, he is not in the least clear where or how they link up with the world of reality. He throws himself on people who cannot understand him, and for him this is one more proof of the abysmal stupidity of man. Or he may develop into a misanthropic bachelor with a childlike heart. He seems prickly, unapproachable, and arrogant; he has a fear of the feminine sex.'

Wolfgang Pauli is full of nervous anticipation the first time he enters Jung's house and climbs the broad, sweeping staircase to the first floor. Dr Jung, a white-haired, fifty-six-year-old man with a pipe wedged in the corner of his mouth, welcomes his new thirty-one-year-old patient. They are in the library, which Jung reserves for sessions with patients he especially wants to engage with. The shelves are filled with books on alchemy, one of Jung's special interests. He invites Pauli to choose one of two easy chairs, the first offering a view of Lake Zurich through the large window, the other facing the bookshelves. C.G. Jung takes a seat on the couch. He can sense the desperate agitation emanating from Pauli. Years later, Jung writes of his first impressions of Pauli:

> He is a highly educated person with an extraordinary development of the intellect, which was, of course, the origin of his trouble; he was just too one-sidedly intellectual and scientific. He has a most remarkable mind and is famous for it. The reason why he consulted me was that he had completely disintegrated on account of this very one-sidedness. It unfortunately happens that such intellectual people pay no attention to their feeling life and so they lose contact with a world that feels and live in a world that thinks: in a world of thoughts merely. So in all his relations to others and to himself he had lost himself entirely. Finally he took to drink and such nonsense so that he grew afraid of himself, could not understand how it had happened, lost his adaptation and was always getting into trouble. This is the reason he made up his mind to consult me.

Pauli begins to talk about his affairs with women, his drinking, his anger, his loneliness, his terribly vivid dreams, his fear of going mad. Jung is fascinated. He is not likely to see such a case, such an intellect, such a test for his own theories about the psyche in his practice again any time soon. What are the forces pulling on this man? How can he be helped to regain some balance? 'And what shall we say of a hard-boiled scientific rationalist, who produced mandalas in his dreams and in his waking fantasies?' asks Jung in his later book *Psyche and Symbol*. 'He had to consult an alienist as he was about to lose his reason because he had suddenly become assailed by the most amazing dreams and visions. When the abovementioned hard-boiled rationalist came to consult me for the first time, he was in such a state of panic that not only he, but I myself, felt the wind blowing over from the lunatic asylum!'

Jung sees Pauli's dream images as archaic symbols, which he recognises from his study of alchemy. He recognises the deep, primal structures of the unconscious mind, which he calls 'archetypes'. They are not delusional, but are a side to Pauli that is struggling to emerge from the shadows. Pauli's fear of women and his aggression towards men are manifestations of his resistance against this dark side. How can so cerebral a person come to terms with such things, immeasurably remote from the concepts of modern physics as they are?

However, Jung is concerned that he, as a strong man, might inhibit the weaker man, Pauli, with his comments or even with his mere presence, and that might prevent him from dreaming or speaking freely. Considering it better to keep Pauli at a distance, Jung decides he should be treated by a female therapist. He hands Pauli a note with the address of his trainee, Erna Rosenbaum, and says goodbye. Rosenbaum has only been working with Jung for nine months. She's a beginner.

Jung believes only a woman can help Pauli to improve his difficult relationship with women in general, and to come to terms with his own feminine, creative side. But Pauli doesn't like the idea at all. He feels insecure and self-conscious around women. That's precisely the point, says Jung, and refuses to take no for an answer. Pauli is disappointed, but he's willing to try anything, and so he reluctantly writes Rosenbaum a rather disgruntled letter requesting an appointment.

Wolfgang Pauli is very distrustful when he goes to see Erna Rosenbaum, and she is horrified by the time he leaves. 'What kind of a man have you sent me?' she asks Jung the next day. 'What's wrong with him? Is he half mad?'

'What happened?' asks Jung.

Rosenbaum tells him that Pauli told her stories that were accompanied by 'such strong emotions that he was thrashing about on the floor'. She asks again, 'Is he insane?'

'No, no,' answers Jung, 'he's a German philosopher. He's not insane.'

Working with Rosenbaum, Pauli starts analysing his dreams. Over the next five months, he records more than a thousand dreams and visions in writing for her. 'I don't envy you, having to read through all that,' he confesses to her. But she does read them, and she is able to help him. His head remains full of dreams and fantasies for the rest of his life, but he is able to pull himself together, avoid sliding back into depression, find a way of dealing with loneliness, and keep his drinking under control.

Jung keeps his distance during Pauli's treatment, but he also keeps himself informed on his progress. From a distance, he observes the 'making of a new centre of personality' in Pauli. He takes Pauli's dream diaries as the basis for his theory on the 'symbols of the individuation process'. Pauli gives his consent, and Jung promises to keep his identity secret, referring in his articles only to 'a scientifically educated younger man'.

Pauli's therapy sessions with Erna Rosenbaum come to an end after five months, and Jung considers it time to take over Pauli's treatment himself.

With time, their doctor–patient relationship develops into a friendship. They discover that they share similar passions. Both are scientists. Both believe physics will always be an incomplete science until it takes into account the influence of the human mind. They talk passionately about science, philosophy, religion, and the history of ideas. Pauli not only joins Jung in his library once a week, but he also often joins the Jung family for dinner.

Jung is obsessed with his idea of archetypes, those primal, unconscious symbols that lie at the base of all our perceptions. He is fascinated by the kabbalah, the ancient, impenetrable tradition of Jewish mysticism.

For his part, Pauli is an admirer of Johannes Kepler, who attempted, unsuccessfully, to derive the structure of the solar system from pure geometry. He also esteems Robert Fludd, one of Kepler's lesser-known contemporaries, who was a member of the secret brotherhood of Rosicrucians, and who believed the key to understanding the cosmos lay in simple geometrical forms. Pauli develops the view that quantum physics must be combined with Jung's analytical psychology to understand the world: the unconscious, the conscious, and everything else as well.

Both men are also obsessed with numbers. Pauli puzzles over the fine-structure constant, that fundamental value of the universe commonly denoted by α, which quantifies the strength of electromagnetic interaction. His teacher, Arnold Sommerfeld, calculated its value to be 1/137. But why 137 in particular? Who or what set α at just the right amount to stop atoms and molecules from collapsing?

One hundred and thirty-seven! Jung is familiar with that number from the kabbalah. Yes, 137 is the numerical value of the kabbalah! Every letter of the Hebrew alphabet is associated with a number, and the sum of the values of the letters in the word 'kabbalah' is 137. Jung and Pauli agree that this can't just be a coincidence. They speculate further. Jung develops his concept of synchronicity to describe circumstances that appear meaningfully related yet lack a causal connection—connections without cause and effect. This is the very issue that plagues quantum physicists. Consciousness and matter, waves and particles, numbers and cosmic order, archetypes and physical theories: somehow, it's all one. The accursed 'Pauli effect' is surely proof that not everything in the physical world can be explained by physics alone.

Jung and Pauli immerse themselves ever more deeply in a world of magic. They author a book together with the title *The Interpretation of Nature and the Psyche*. It is scientifically worthless, but for Pauli it is therapeutic. In a letter to his friend and assistant Ralph Kronig, he ascribes his recovery to an 'acquaintance with psychical things, of which I was not aware before and which I will summarise under the name "the autonomous activity of the soul"'. Pauli believes in 'spontaneous growth processes' and 'something objectively psychical, which cannot be explained with material causes'. He signs that letter 'Your old and new W. Pauli'.

At the end of October 1934, Pauli ends his psychoanalysis by Jung, but the two men remain penfriends, and Pauli's dreams continue to provide Jung with research material. Pauli has now recovered enough to marry again. And this time, the marriage lasts. He and his wife, Franca, lead a bourgeois life, but have no children. On 5 December 1958, Pauli is hospitalised with severe stomach pains. When he sees his room number at the Red Cross hospital, he exclaims 'It's 137! I won't be leaving this room alive.' Ten days later, Wolfgang Pauli is dead.

COPENHAGEN, 1932

Faust in Copenhagen

Copenhagen, spring 1932. The quantum physicists meet at Niels Bohr's house every year over Easter. The colleagues congregate to spend a pleasant week together eating, making music, hiking, swimming, and discussing their favourite topics.

Werner Heisenberg travels from Leipzig, Paul Dirac comes from Cambridge, and Paul Ehrenfest arrives from Leiden. Lise Meitner, the 'German Marie Curie' (according to Albert Einstein), reports to the group on her experiments with radioactive elements. Bohr tells the guests to feel free to bring their talented students along. Werner Heisenberg brings his twenty-year-old doctoral student Carl Friedrich von Weizsäcker, a fine-mannered diplomat's son who will soon become Heisenberg's best friend. Weizsäcker introduces the politically inexperienced Heisenberg to a world beyond physics. They go skiing together, just as Heisenberg had once done with his mentors Sommerfeld and Bohr. 'Only the friendship with Carl Friedrich, who struggles in his own serious way with the world around us, leaves open to me a small entry into that otherwise foreign territory,' Heisenberg will later write in a letter to his mother.

These meetings in Copenhagen have a strong influence on Weizsäcker's development as a scientist. He's able to experience at first hand Paul Dirac outlining his proposed anti-electrons. And to listen to Niels Bohr and Paul Ehrenfest discussing the minutiae of interpretations of quantum mechanics. At this year's meeting, Wolfgang Pauli is responsible for the number-one topic of discussion: his neutrino hypothesis. Nineteen thirty-two will later become known as a miracle year for experimental physics. Working at the Cavendish Laboratory in Cambridge, Rutherford's student James Chadwick discovers the neutron. At Caltech in California, Carl D. Anderson passes cosmic rays through a cloud chamber and discovers the anti-particle to the electron, calling it the positron. At the same time, researchers in Berkeley, California, start working with particle accelerators.

Physics research is increasingly becoming a matter for big teams using large apparatus. Solitary geniuses working at their desk with just paper, a pencil, and their mind are becoming the exception. Both theory and experimentation remain cornerstones of physics research, but experimentation now increasingly takes centre stage over theoretical physics. Carl Friedrich von Weizsäcker's research will differ from that of his teachers. Great breakthroughs are no longer made in the mind, or during walks on the beaches of Heligoland, or at Christmas parties in Arosa, or at meetings like this one in Copenhagen. They are made at large-scale research facilities. Physics is becoming a practical discipline—and the consequences of this will be huge.

Wolfgang Pauli is not in attendance in Copenhagen this year. He's struggling with a crisis in his life. Albert Einstein never attends, despite his friendship with Bohr. He's not interested in the concerns of these young scientists—with their nuclear physics and such stuff. Teaching doesn't interest him. He has no doctoral students.

There is no official agenda in Copenhagen, but there is a tradition. The young physicists perform a sketch making fun of the old guard—some of whom are not even thirty years old yet. The previous year they had written a spoof of a spy film that many of the physicists saw together at the cinema.

This year, the young physicists choose a more ambitious subject. It's 100 years since Johann Wolfgang von Goethe died, on 22 March 1832. The scientists take this as inspiration for a parody of Goethe's most famous drama, *Faust*. Centenary celebrations are being held all over Germany. Goethe is an integral part of every school syllabus in the country, and students often have to learn his verses by heart. Even Niels Bohr's father could recite Goethe from memory. The choice of subject is also a nod to Werner Heisenberg, who is a great fan of Goethe, both as a poet and a scientist.

Max Delbrück, a brilliant twenty-five-year-old student from Berlin, writes the script. He had planned to become an astronomer until he happened to attend a talk by Werner Heisenberg in 1926. Despite not understanding any of it, he was fascinated by what he heard, and immediately began studying atomic physics under Wolfgang Pauli, Max Born, and Niels Bohr. The previous year, at a party in Rome, he had a flirtation with Ève Curie, who was accompanying her mother, Marie, on a trip around Europe, but Ève rebuffed him. His pride injured, Delbrück soon abandoned physics to study biology. He will research the quantum physical foundations of life and become the founding father of molecular biology, in much the same way that Niels Bohr is the founding father of quantum physics.

George Gamow, Bohr's sharp-witted, twenty-eight-year-old student from Odessa, draws the illustration for the script. It was Gamow who came up with the idea for the spy-film spoof the previous year. This year, he's not present at the gathering, as the Soviet state has refused to issue him with a passport. Stalin, who is currently responsible for great suffering in Gamow's homeland of Ukraine, due to the famine he's caused and his political purges, wants to prevent Soviet scientists from 'fraternising with scientists from capitalist countries'. Gamow will seize the next opportunity to flee to the West.

Delbrück calls his sketch 'Faust in Copenhagen'. The roles in Goethe's play are assigned a double meaning. Paul Ehrenfest is Faust, and Wolfgang Pauli plays Mephisto. Niels Bohr takes the part of God, and Gretchen is represented by the neutrino. The performance takes place at the end of the Easter gathering, in the lecture theatre of Bohr's institute. There are no props: only a lectern, a bench, and a couple of chairs.

Léon Rosenfeld, Bohr's twenty-seven-year-old assistant, plays Wolfgang Pauli, alias Mephisto, the troublemaker who tries to lead Paul Ehrenfest—that is, Faust—astray. Rosenfeld has a similar appearance to Pauli: small, plump, bald.

Felix Bloch, a twenty-six-year-old who is Heisenberg's first doctoral student, is currently a visiting researcher in Copenhagen and living in the attic flat where Heisenberg discovered the uncertainty principle. He takes the part of Niels Bohr—that is, God. Bloch is fit and well built.

Ehrenfest doesn't think much of Pauli's neutrino proposal. So Delbrück has his Faust explain that the idea of a particle with neither mass nor charge is pure madness. Such a thing cannot exist. Simply impossible!

Ehrenfest also has doubts about the version of quantum electrodynamics that Pauli and Heisenberg have been working on for years. They are grappling with the idea that the values for the mass and charge of an electron tend towards infinity, and they have not yet found a way to rein them in. What is the point of these fancy mathematical gimmicks, Ehrenfest asks his colleagues. He has no time for this new fashion of developing physical hypotheses from aesthetic considerations. Beauty is a matter for tailors, he says, not scientists.

Goethe's *Faust* opens with a prologue in which the archangels Raphael, Michael, and Gabriel praise the Lord for his creation. Mephisto mocks their adulation. Delbrück turns the archangels into three astrophysicists who enthuse about how well physics describes the world: solar radiation, the splendour of binary stars. Mephisto, complete with horns and a devil's tail, bursts in on their conversation and curses, 'The entire theory is crap.' On the stage sits a figure covered in a white sheet. It rises, and Felix Bloch reveals himself, dressed unmistakably as Niels Bohr. God asks Mephisto:

Hast thou, then, nothing more to mention,
Com'st ever thus with ill intention?
Find'st nothing right in physics, eternally?

To which Mephisto answers:

No, nonsense! I find things still bad as they can be.
It e'en to pity moves my nature,
To plague the physicist, the wretched creature.

The issue is the neutrino, the bone of contention between Pauli and Bohr. Pauli postulated the existence of the neutrino to counter Bohr's theory that the fundamental law of the conservation of momentum and energy is valid only as a statistical mean in some processes at the atomic level.

God dismisses Mephisto's neutrino hypothesis with Bohr's dreaded phrase 'That's very interesting! … But, but …' Mephisto responds in the style of Pauli: 'No, shut up! Stop it, that's nonsense!'—'But Pauli, Pauli, we are more in agreement than you think!' Laughter in the audience. They recognise these phrases from discussions between Bohr and Pauli. Bohr laughs along with the rest. Unlike Goethe's God, he can laugh at himself.

Bloch, alias God, and Rosenfeld, alias Mephisto, make a bet over the soul of Faust, that is, Ehrenfest, the 'schoolmaster'. In Goethe's play, Mephisto gives Faust a potion. Faust falls in love with Gretchen and seduces her.

In Delbrück's play, Mephisto disguises himself as a travelling salesman, complete with a bowler hat on his head. He tries to sell Faust some quantum electrodynamics 'from Heisenberg-Pauli'. 'I don't like it!' cries Faust. 'Perhaps some from Dirac? With infinite self-energy?'—'I don't like it!' cries Faust again. Mephisto still has 'something unusual' on offer: the neutrino. Faust refuses that, too:

> Thou shalt nevermore seduce me.
> When I say to the theory flying
> Linger a while, thou art so fair
> Then bind me in thy bonds undying
> And my final ruin I will bear!

Then the neutrino itself enters, as a singing Gretchen:

> My charge is gone,
> Statistic's sore,
> I never shall find it,
> Ah, nevermore!

Save you have me near,
No equation will appear,
The world is gall
And bitterness all.

But Ehrenfest remains steadfast against the charms of the neutrino. With the line 'Ehrenfest, I shudder at thee!', Gretchen exits.

The sketch grows increasingly absurd. The scene in Auerbach's cellar is transposed to a bar in Ann Arbor, in the US, where Pauli broke his shoulder falling down the stairs two years earlier. Einstein is extolled as 'The King'; his 'new field theory' is mocked as a flea. The stage becomes the scene of a 'classical' and a 'quantum theoretical' Walpurgis Night. The saviour appears in the form of the neutron, whose existence was proven just two months earlier. 'The eternally neutral draws us heavenward!' sings the Chorus Mysticus at the end of the play.

In the years that follow, the Copenhagen gatherings continue, as does the sketch tradition. Until the Second World War puts an abrupt end to all such merriment.

BERLIN, 1933

Some flee; some stay

On 30 January 1933, Adolf Hitler is appointed Reich chancellor of Germany. The later propaganda minister, Joseph Goebbels, writes in his diary that evening, 'We all have tears in our eyes. We shake Hitler by the hand. He deserves it. Great jubilation. And the people down there are running riot. Time to get to work. The Reichstag will be dissolved.' Hitler, Goebbels, and company don't waste any time. In the weeks that follow, they turn Germany into a Nazi state. They abolish freedom of the press and freedom of assembly, re-arm the Wehrmacht, and pass the Enabling Act, which ends parliamentary democracy. Many people fear Germany will turn into a dictatorship. 'Nonsense,' says Wolfgang Pauli. That's the kind of thing that happens in Russia, but not in Germany—impossible!

For the half a million Jews who live in Germany, it's the dawn of a very dark period. The exodus begins slowly. Only 25,000 have left the country by June 1933. As violence against Jews increases under the Nazis' encouragement, 17 million Germans give their votes to the Nazi Party in the elections of 5 March 1933.

On 23 March 1933, the Reichstag passes the Enabling Act, formally known as a 'Law to Remedy the Distress of People and the Reich', gaining the necessary two-thirds majority with the support of the Nazi Party, the right-wing German National People's Party, the Catholic Centre Party, and other conservative groups. Only the much-diminished Social Democratic Party resists the pressure to vote for the law. Its chairman, Otto Wels, says in a speech, 'You can take our freedom and our lives, but you cannot take our honour. No

Enabling Act gives you the power to destroy ideas that are eternal and indestructible.' Parliamentarians from the German Communist Party are excluded from the vote—because they are either in prison or in hiding. SA stormtroopers intimidate any members of parliament they suspect might vote against the Act—both outside and inside the chamber.

Albert Einstein declares he will not return to Germany after his current lecture tour of the US. 'It takes a great deal of courage these days to say and do the obvious thing, and there are truly very few who show such courage. You are one of those few and I shake your hand as someone who is close to me in all senses,' writes Einstein in a letter to the Viennese physicist Hans Thirring. 'We see with frightening clarity that we must fight and that we must convince those who are still decent that they must not stand aside.' The brutality of the Nazis moves Einstein to rethink his strictly pacifist views. 'I loathe all armies and any kind of violence, yet I am firmly convinced that, in the present world situation, these hateful weapons offer the only effective protection.'

On 7 April 1933, the Nazis pass a 'Law for the Restoration of the Professional Civil Service'. It affects two million public-sector employees, targeting the Nazis' political opponents, including socialists and communists, as well as Jews. Paragraph 3 of the law is an example of an 'Aryan paragraph': 'Civil servants of non-Aryan descent are to be retired; if they are honorary officials, they are to be dismissed from their official status.'

Anyone with at least one Jewish parent or grandparent must be fired or forcibly retired. Reich president Hindenburg manages to persuade Hitler to allow an exception for 'Non-Aryan' civil servants who fought for Germany or its allies on the frontline in the First World War, as well as those whose fathers or sons died in active service. All civil servants are obliged to 'prove' they are 'Aryan' within two weeks, by providing their birth, baptism, and marriage certificates, and those of their parents and grandparents.

The Imperial Constitution of 1871 had awarded equal status to Jewish citizens of the empire. Now they face legalised state discrimination. And the 'Aryan paragraphs' are only the beginning.

Universities, as state institutions, also fall under the Law for the Restoration of the Professional Civil Service. The president of the Kaiser Wilhelm Society, Max Planck, is not a Nazi. But he does the Nazis' bidding. He writes in a telegraph to Hitler, 'The members gathered for the 22nd Ordinary General Assembly of the Kaiser Wilhelm Society for the Advancement of Science are honoured to send respectful greetings to the Reich Chancellor, thereby taking the opportunity to solemnly pledge the utmost efforts of German science to aid in the reconstruction of the new nation state, which wishes to be its protector and patron.'

More than a thousand academics, including 313 professors, are removed from their positions. Almost a quarter of Germany's university-level physicists are forced into exile, including nearly half of the country's theoretical physicists. By 1936, more than 1,600 scholars have been expelled, a third of whom are scientists, including twenty current or future Nobel Prize winners: eleven in physics, four in chemistry, and five in medicine. The Jewish banker and patron of German science, Leopold Koppel, is forced off the committee of the Kaiser Wilhelm Institute, for which he provided the funds.

Max Planck can't stand idly by and watch as this exodus happen. He seeks an audience with Hitler to try to limit the damage to German science. He's granted it on 16 May 1933, at 11.00 am. One must differentiate, Planck tells Hitler. There are different kinds of Jews: those who are 'valuable to mankind' and those who are 'worthless'. He gives the example of Fritz Haber, the Christianised son of Jewish parents and the winner of the Nobel Prize for Chemistry for his invention of a method to synthesise ammonia for fertilisers, and whose work on chlorine gas led to the use of chemical gas as a weapon in the First World War, thereby serving the German fatherland. But Hitler is not interested in any such differentiation, crying, 'A Jew is a Jew. Jews all stick together like limpets.'

'But it would be an act of absolute self-mutilation,' argues Planck, 'to force "valuable" Jews to emigrate. That would lead to a lack of scientific expertise in Germany and benefit foreign countries.'

Hitler flies into one of his much-feared rages, thumping his knee as he spews words faster and faster, screaming at the seventy-five-year-old professor and threatening to send him to the concentration camps. Planck can only listen in silence before slinking away. 'Pitiable muddle-head,' Hitler shouts after him as he leaves.

Planck knows what it's like to make sacrifices. He had done much to help ailing German research recover after the First World War. He lost his elder son in the war. But such senseless sacrifice? He can't see any advantage to that. However, after this disastrous conversation, Planck realises that Hitler would rather sacrifice German science than give the impression of showing leniency towards Jews.

In April 1933, Pascual Jordan, once the stammering student who helped formalise matrix mechanics, becomes a member of the Nazi Party. And at the beginning of May, Johannes Stark is named president of the Imperial Physical-Technical Institute (PTR). The Nobel laureate is a dogmatic and belligerent anti-Semite who has supported Hitler since the 1920s. When Hitler was imprisoned in 1924, Stark and Philipp Lenard publicly expressed their support for him in a newspaper, praising 'that same spirit of total clarity, of honesty towards the outer world, yet inner consistency, which we recognised and revered long ago in the great researchers of the past, in Galilei, Kepler, Newton, Faraday, we admire and revere in Hitler, Ludendorff, Pöhner and their comrades; we recognise in them our closest kindred spirits'.

Now, ten years later, Stark is out for revenge. In the 1920s, he had resigned his professorship in Würzburg to set himself up in business — first as a porcelain manufacturer, then running a brickworks. When his business ventures failed, he tried to re-enter academia, but no one was willing to give him a position. Now he sees the opportunity to exact revenge on all those who spurned him. Together with his fellow physicist, the avowed Nazi Philipp Lenard, he sets out to establish 'Aryan physics'. As president of the PTR, he is in charge of distributing research funds. He wants to be the leader — or Führer of German physics, just as Hitler is the Führer of the people.

On 10 May 1933, book burnings take place across the German

Reich. They are organised by the German Student Association's Central Office for Press and Propaganda, in twenty-two locations, mostly university towns such as Bonn, Göttingen, and Würzburg, all following the same script. SA stormtroopers and party members bearing burning torches and waving swastika flags march to a public square and toss 'Jewish', 'Bolshevist', and 'un-German' books onto the fire—in a prescribed order, sorted by author. They all recite the same 'fire oaths', scripted by the Central Office. In Munich, books are burned on Königsplatz. The burning in Berlin is on a particularly large scale—40,000 people gather in the square between the university and the opera house to watch the satanic spectacle. For a distance of eight kilometres, students escort the convoy of trucks and cars laden with 20,000 books pillaged from libraries and bookshops. Student fraternities turn out in their blue or purple uniforms, with red or green caps, white breeches, and high boots with spurs. They sing both Nazi anthems and student songs. The square is prepared with a thick layer of sand, on which the books are piled several metres high. The students set the pyres alight with their torches as they march past. The names of the authors to be erased are chanted as their books go up in flames.

'No to the soul-fraying overestimation of a life of sexual urges, yes to the noblesse of the human soul! I consign the writings of Sigmund Freud to the flames.'

'No to literary betrayal of the soldiers of the world war, yes to the education of the people in the spirit of truthfulness! I consign the writings of Erich Maria Remarque to the flames.'

'No to opportunism and political betrayal, yes to dedication to the people and the state! I consign the writings of Friedrich Wilhelm Förster to the flames.'

The works of Albert Einstein also end up on the bonfires, along with those of Karl Marx, Bertolt Brecht, Émile Zola, Marcel Proust, and Franz Kafka. Around midnight, Goebbels mounts the speaker's dais and declares, 'Jewish intellectualism is dead. The German soul can express itself once more.' Clever people like Max Planck cannot miss these warning signs, but they do nothing. Adolf Hitler has now been in power for precisely one hundred days.

When reports of the events in Germany spread around the world, scientists and scientific associations begin organising support, with both money and job offers, for their colleagues forced to flee Nazi oppression. Aid organisations are set up, financed by donations. The Academic Assistance Council (AAC) is founded in London in April 1933, with Ernest Rutherford as its chairman, to help refugee scientists, artists, and authors. Many of them initially escape Germany by crossing the border into Switzerland, the Netherlands, or France, and, after a short stay, travel on to the United Kingdom or the United States.

Despite the race laws forcing excellent scientists out of state jobs, and the mass exodus or deportation of researchers, many competent physicists remain in Germany. Otto Hahn, the man who discovered nuclear fission, stubbornly refuses to leave, and is, in the words of Albert Einstein, 'one of the few who retained their decency and did their best during those terrible years'. As he continues his research, Hahn, along with his wife, Edith, advocates for Jewish colleagues who are under threat, and corresponds with others who are in exile, including his collaborator Lise Meitner. Hahn's friend and colleague Max von Laue, who was awarded the Nobel Prize in 1914, goes even further, putting his life in danger by repeatedly and publicly criticising and condemning the Hitler regime.

Physics comes to a halt while physicists are in turmoil. There are more important things these days than matrices and wave-particle dualities. Some physicists are busy developing sonar devices for U-boats; others are building code-cracking machines. In Copenhagen, Niels Bohr turns his institute into a base for stranded physicists. In April 1934, he organises a guest professorship for James Franck, Max Born's Göttingen colleague, who is from a Jewish family. After a year in Denmark, Franck moves on to the United States to take up a professorship in physical chemistry in Chicago.

In December 1931, the Royal Danish Academy of Sciences named Bohr as the next occupant of the *Æresbolig*, the 'honorary residence' built by the heir to the Carlsberg brewing fortune. This is seen as recognition of Bohr, not as Denmark's most important physicist, but as the country's most important citizen. Bohr uses his influence at home and abroad to help others. Both Niels Bohr and his brother Harald are instrumental in establishing the 'Danish Committee for the Support of Intellectual Refugees'.

Max Born is also from a Jewish family. Until the present time, he says, he 'never felt particularly Jewish', but now he is 'extremely conscious of it, not only because we are considered to be so, but because oppression and injustice provoke me to anger and resistance'. Due to his service as a wireless operator in the First World War, he has the right to claim exemption from the 'Aryan paragraph'. But he prefers not to make use of that right, believing that to do so would be going along with the Nazis. On opening the local paper one day, he discovers his name on a list of civil servants who are to be suspended from their jobs, alongside the names of the mathematicians Richard Courant and Emmy Noether. The following evening, the Born household starts receiving threatening telephone calls.

Born takes long walks through the forest, thinking and trying to decide what to do. Where should he go? One thing is clear: carrying on as before is not an option. He writes to Ehrenfest, 'I shudder when I think I might for some reason have to face the students again who threw me out, or if I had to live among "colleagues" who accepted it so easily.'

He gradually reaches the conclusion that he and his family must leave Göttingen, the city where he has spent most of his adult life, in which he created a centre for the study of physics and in which his wife, Hedi, was born and gave birth to their son, Gustav. With only light luggage, they board a train on 15 May. 'Everything I had built up in twelve years in Göttingen was destroyed,' Born will write later. 'It felt like the end of the world.'

The Borns go to Val Gardena in the Dolomite Mountains of South Tyrol. They find lodgings with a local farming and wood-carving family by the name of Perathoner, in the village of Wolkenstein, called Sëlva in the local Ladin language. Spring is on its way, the sun rising ever higher over the rugged peaks of the Dolomites, and the snow begins to melt. Hermann Weyl pays them a visit in Wolkenstein, and is soon joined by his lover, Anny Schrödinger, who has been staying nearby with her husband, Erwin, to recover from the recent horrors of Berlin. Wolfgang Pauli arrives next, with his sister Hertha, who is a journalist and actress but has lost all opportunity to work in Berlin under the Nazis. Two of Born's students seek him out in Wolkenstein and join him. Born lectures them from a bench outside the house or in the middle of the forest. He's proud of his 'little Sëlva University'. There is an idyllic feeling about this place and time, albeit a fleeting one.

One person who does not turn up is Werner Heisenberg. He pens a letter to Born in Wolkenstein to convince him to return to Germany, assuring him he will be safe there because 'only a very few [Jews] are affected by the law'. There's no doubt about it, he says, because Hitler gave Max Planck a promise.

Heisenberg will later say that 1933 spelled the end of 'the golden age of atomic physics'. The short few years when scientists were free to ponder blithely on atoms, electrons, waves, and matrices are now over. Darkness is rising, and Heisenberg can see that. Friends notice that the normally jaunty glint in his eye is gone. He increasingly withdraws into himself.

Heisenberg tries to accommodate himself to the regime. He remains a state employee of the Third Reich. Like all 'Aryan' professors who stay, including Planck, he has to submit his parents' birth and marriage certificates for verification, is forced to attend Nazi indoctrination camps, to swear an oath of allegiance to Hitler, and to begin all his lectures with '*Heil* Hitler'. He considers resigning his professorship in protest, and seeks Planck's advice on the matter. Planck tells him it would be pointless, and would only free up his professorship to be taken by a staunch Nazi and bad physicist. Planck and Heisenberg continue to hope that there will be a political revolution that will not adversely affect science, and that the whole situation will eventually calm itself.

But the damage has already been done. Within just a few weeks, the University of Göttingen, the place where Carl Friedrich Gauss, Georg Christoph Lichtenberg, and David Hilbert all worked, the cradle of quantum mechanics, is transformed into a second-rate college by the Nazis. When the celebrated mathematician David Hilbert is asked in 1934 by the Reich minister for science, schooling and people's education, Bernhard Rust, whether it was true that his institute has suffered 'under the departure of the Jews and Jew-friends', Hilbert simply answers, 'The institute — no longer exists.'

On 13 December 1935, the Physics Institute at the University of Heidelberg is officially renamed the Philipp Lenard Institute at a ceremony that minister Rust is unable to attend due to illness. Now the institute's director, Lenard has recently published a book called *German Physics in Four Volumes*. In it, Lenard claims that science is a matter of race, of blood, like the whole of human endeavour. He writes that the Jews have their own physics, which is totally different from 'German', 'Aryan', or 'Nordic' physics. Lenard is not able to give an explanation of the nature or origin of that difference in the four volumes of his work. 'Jewish' physics is whatever he and his ideological bedfellows say it is. In his speech at the renaming ceremony, Johannes Stark, president of the Imperial Institute and one of Lenard's ideological bedfellows, rails against supporters of Albert Einstein, and bemoans the fact that Max Planck is still the chairman of the Kaiser Wilhelm Society. The ceremony ends with shouts of '*Sieg Heil*' and a rendition of the Nazi anthem *Raise the Flag*.

Max Born is less naive than Heisenberg. He plans to take his family to England, because 'the English seem to administer to the displaced in the most generous and noble manner'. Born has been suspended from work, but he is free to travel. He is one of the very few German scientists whose fame is so widespread that they are recognised all over the world.

Cambridge University offers him a three-year teaching contract, which he accepts after ensuring that the post has been created especially for him, as he doesn't want to deprive a British scientist of a job. This is the start of a restless time for the family. When Born's tenure at Cambridge is up, he accepts the offer of a guest lectureship at the Indian Institute of Sciences in Bangalore. But the institute's management doesn't consider Born's theoretical research to be of any practical use, and doesn't offer him a permanent position. After just six months, the family has to move on again. While Born is considering taking a job in Moscow, he receives an offer to become the next Tait Professor of Natural Philosophy at the University of Edinburgh in 1936.

The horrors of the Second World War weigh heavily on him, even though he lives in physical safety. He's convinced that Adolf Hitler must be defeated, but how? Born knows that violence is the only way. At the same time, he's appalled by the Allies' aerial bombing campaigns, which reduce entire cities to rubble. 'The idea that Hitler could be overthrown by killing women and children and destroying their homes seemed absurd and abhorrent to me,' he will later write. Towards the end of the war, Max Born suffers such severe depression that he's unable to carry on working. The war has destroyed everything. The world has gone fuzzy, and not just for quantum physicists. The contours of everything that was familiar, everything that was certain, are beginning to blur.

AMSTERDAM, 1933

A sad end

Leiden, 1933. Paul Ehrenfest watches with fear and horror as Adolf Hitler rises to power and German science falls apart. He's losing friends, one by one. Hendrik Lorentz is dead. Max Planck is sinking ever deeper under the weight of having to represent his country's science community despite disapproving of its actions. Albert Einstein, with whom Ehrenfest shared some of the happiest moments of his life, has fled the Nazis and gone abroad. That's the most painful loss for Ehrenfest. He suspects he'll never see Einstein again.

Teaching and research no longer provide Ehrenfest with any comfort, although he had been one of the most respected theoreticians in the world in his younger years. He had been Hendrik Lorentz's personal choice to succeed him as professor in Leiden. But for quite some time now, he has no longer been able to find the energy to follow the latest developments in atomic physics.

Paul Ehrenfest was born in Vienna in 1880 and studied under the great Ludwig Boltzmann. He and his wife, the Russian mathematician Tatyana Ehrenfest-Afanasyeva, made a number of important contributions to the field of statistical mechanics, as the couple moved around Europe: Vienna, Göttingen, and St Petersburg, always on the verge of poverty. Until Ehrenfest was appointed physics professor at Leiden University in 1912, on the strength of a recommendation from Albert Einstein, who himself preferred to move to Zurich.

Ehrenfest turns Leiden into a major centre of theoretical physics. His good friend Einstein calls him 'the best teacher of our subject that I have ever met'. Whenever the hustle and bustle of Berlin becomes too much for him, Einstein escapes to Leiden to enjoy the warm, affectionate atmosphere of the Ehrenfest household. Paul and Tatyana don't smoke or drink, but they allow Einstein to light up his pipe in the guest room.

With his wit and kind heart, Ehrenfest is one of those rare people who are liked by everybody. He first meets Wolfgang Pauli at the 'Bohr Festival' in 1922. Both had written an entry for the *Encyclopaedia of Mathematical Sciences*, and they share a similar sense of humour. 'Herr Pauli, I like your encyclopaedia entry even more than I like you!' jokes Ehrenfest. 'How funny, I find the exact opposite is the case with you!' retorts Pauli. It's the start of a close friendship and a 'war of witticisms' waged by the two friends in person and by letter. In his letters, Ehrenfest addresses Pauli as 'My dear, dreadful Pauli' or 'St Pauli', in reference to Pauli's habit of hanging out in Hamburg's famous red-light and drinking quarter. He also calls his new young friend 'the scourge of God', which Pauli is so proud of that he takes to signing off in his letters to Ehrenfest with 'All the best, from SG'. Ehrenfest signs his letters to Pauli 'Your devoted Schoolmaster'.

However, unnoticed by almost everyone, something is brewing inside Paul Ehrenfest. In 1931, he writes in a letter to Niels Bohr that he feels left behind by the rapidly advancing new young physicists: 'I have lost all contact with theoretical physics.' In a letter dated 15 August 1932 and addressed to seven 'dear friends', including Bohr and Einstein, he writes of his 'weariness of life' and his decision to 'vacate his position in Leiden'. 'My interest in understanding the advances in physics and my great joy in imparting that understanding to others were actually the backbone of my life. Finally, after increasingly nervous and ever-more ragged attempts, I have given up in despair. This is an incurable, devastating sickness at the very core of my being.' He never sends the letter.

Ehrenfest is trying to find a new place for himself in physics. In summer 1932, he publishes an article in the *Zeitschrift für Physik* with the title 'Some exploratory questions regarding quantum mechanics'. He writes to Pauli that he had 'been choking on the decision' to publish for more than six months, until he was eventually driven to do it by 'a kind of desperation'. He feels helpless, as the 'schoolmaster' of his students, who expect him 'to know and understand everything'. In this letter to Pauli, he pours out his heart about the aspects of quantum mechanics that plague him. Pauli responds like a true friend, writing a long letter saying he's always happy to read Ehrenfest's questions, and will try to answer them as best he can.

Pauli's response has a positive effect on Ehrenfest, but it doesn't last long. Ehrenfest has a hard time coming to terms with the fact that his now-sixteen-year-old son, Vassily, has Down syndrome. His marriage to Tatyana breaks down, and in autumn 1932 she returns to Russia to take a job in a small town in the Caucasus Mountains. Ehrenfest considers joining her there, and travels to Russia at the end of the year. But, after experiencing the terrible shortages people are suffering there, he decides not to follow her.

At Easter 1933, he writes a letter to his former student Hendrik Casimir, who is now Pauli's assistant in Zurich: 'Oh, Caasje, place your broad shoulders beneath the cart of physics in Leiden.' Casimir and Pauli are both at a loss. What does Ehrenfest mean? He wants to secure a successor for the professorship in Leiden, just as Hendrik Lorentz had done two decades earlier.

In May 1933, Ehrenfest travels to Berlin. He visits Max Planck. Planck has changed. 'As Planck was talking to me, I could see how the man was suffering. In those days I met no one more desirous of the release of death,' Ehrenfest writes to Einstein on 10 May 1933. 'The man' he was writing about was himself.

On 25 September 1933, Ehrenfest visits his son, Vassily, in an Amsterdam hospital. Not wanting Vassily to be left without care after his death, Ehrenfest shoots him. Then he turns the gun around and kills himself. Vassily survives, but is left blind in one eye.

Albert Einstein is in Norfolk when he receives news of his friend's death, but he has no time to mourn. He fears for his own life, knowing the Nazis' bloodhounds are after him.

OXFORD, 1935

The cat that isn't there

Oxford, 4 November 1933. Erwin and Anny Schrödinger arrive in the English university town where they are hoping to stay. They want to get away from the Nazis, who have made life in Berlin unbearable for them. Erwin Schrödinger has a fellowship at Oxford's Magdalen College. He and Anny rent a large house at 24 Northmoor Road. One of Schrödinger's friend the South Tyrolean physicist Arthur March, and his wife, Hilde, move into a smaller house at 86 Victoria Road, only twenty minutes' walk away. Schrödinger has organised a visiting professorship at Oxford for March. Hilde is pregnant, but the baby's father is Schrödinger.

Schrödinger and women: that's a broad subject, and one that is almost as complicated as quantum mechanics. Schrödinger had spent summer 1932 in Berlin with Itha Junger, whose private maths tutor he had been when she was a vivacious fourteen-year-old girl. She was now a beautiful twenty-one-year-old woman, and has been his faithful lover for four years. As autumn approached, she found out that she was pregnant. Schrödinger reacted to the situation in his typical way, ending the affair, only to start a new one, this time with Hilde March. He had taken Arthur March on as his assistant just to get closer to Hilde. While Anny was spending the summer of 1933 visiting Max Born in Wolkenstein with Hermann Weyl, her husband was on a cycling holiday with Hilde March.

At the same time, Schrödinger longed for a child, and hoped that Itha would give him a son. She did not comply with his wish, terminating her pregnancy and leaving Berlin to live far from

Erwin Schrödinger, one of the founding fathers of wave mechanics, 1933

the cities where she had grown up and had settled down with Schrödinger. Following the injuries she suffered due to her breakup with Schrödinger, she had a number of miscarriages, and was to remain childless for the rest of her life. Hilde March, on the other hand, has been carrying Schrödinger's baby since their summer bike tour.

In September, Schrödinger bumps into Hansi Bauer-Bohm in a grocery shop on Lake Garda. He had known her when she was a young girl in Vienna. Now twenty-six, she is in Garda on her honeymoon, but is already unhappy in her marriage. Another affair of Schrödinger's takes its course.

The Schrödingers have barely arrived in Oxford when Erwin is notified that he is to receive the 1933 Nobel Prize for Physics—together with Paul Dirac. It seems that a new and better phase of Erwin's life is starting. The award ceremony takes place in Stockholm on 10 December, the anniversary of Alfred Nobel's death. Erwin and Anny pull in to Stockholm's central station on 8 December. Dirac arrives from Cambridge accompanied by his mother; and Werner Heisenberg, who is to receive his 1932 Nobel Prize belatedly, travels with his mother from Leipzig. Erwin Schrödinger closes his speech at the banquet with the words, 'I hope to return here sometime soon—and not just once, but many times. When I do, it probably won't be to attend banquets in halls festooned with flags, and I won't need to pack so many formal clothes, but rather with a pair of skis across my shoulders and a rucksack on my back.' Paul Dirac gives a rambling speech about the similarities between atomic physics and the world's current economic problems. Werner Heisenberg does not deliver a speech, simply thanking his hosts for their hospitality. Schrödinger deposits his prize money of 100,000 crowns in a Swedish bank, just to be on the safe side. This year, no Nobel Peace Prize is awarded.

Schrödinger's arrival in Oxford with his entourage of women causes consternation. By the same token, Schrödinger also feels alienated by the customs that prevail at Oxford. He calls the colleges 'universities of homosexuality' in a letter to Max Born. Women are alien creatures in this world of all-male professors' dinners, while Schrödinger sets up home with two such 'aliens'. He strolls through Oxford with his pregnant mistress, Hilde March, making no secret of their relationship. He feels the disapproving glares of the people they pass, and will never feel at home in Oxford.

Hilde March gives birth to her and Schrödinger's baby: a daughter whom they name Ruth. She's the eldest of three daughters Schrödinger will father with three different lovers. The Marches and the Schrödingers care for Ruth together. However, the social stigma of being Schrödinger's 'second wife', and the mother of his daughter, take their toll on Hilde. Arthur March's two-year tenure as visiting professor in Oxford ends in 1935. He and Hilde return to Innsbruck with little Ruth, and Hilde spends several months in a sanatorium recuperating from the stresses of the past few years. Schrödinger's tenure is extended for a further two years, and he stays in Oxford.

Hansi Bauer-Bohm and her husband, Franz—who are both of Jewish descent—have now fled the Nazis and are in London. That's a happy coincidence for Schrödinger: he and Anny also have a flat in London, which was intended for Anny to use while her husband spent time alone with Hilde in Oxford. Now, Erwin uses the flat to be alone with Hansi. He also goes on holiday with her in the summer of 1935.

It's astonishing that Schrödinger still has any time to devote to physics at all. He engages in a deep discussion by letter with Albert Einstein about the foundation of quantum mechanics. The two agree that the foundation is weak. Schrödinger and Einstein are the last great dissidents of quantum mechanics. 'Dear Schrödinger, you are practically the only person with whom I really like to argue,' Einstein writes from Princeton, 'because all the other fellows do not view the theory from the facts, but only view the facts from the theory.'

Schrödinger has become homeless, in both a geographical and a scientific sense. He was one of the driving forces behind the development of quantum mechanics, but now it has become foreign to him. 'Whenever I talk to the highly gifted young people here,' he writes to Ehrenfest from Berlin, 'I always, always have the feeling that they don't understand what I find so utterly unbearable about the lush forest of theories that has been shooting up in the past few years.' Einstein does understand it. Together they lay bare what it is about quantum mechanics that they so dislike.

In a letter to Schrödinger, Einstein attempts to illustrate 'the evils we see' with an example. An imaginary 'pile of gunpowder that, due to some inner forces, has a mean probability of exploding in a year'. In the course of that year, Einstein continues, the pile of gunpowder's 'ψ-function'—that is, its wave function—is therefore 'a sort of blend of not-yet and already-exploded systems'. Einstein finds that absurd. 'In reality there is no intermediary between exploded and not-exploded.'

Inspired by his correspondence with Einstein, Schrödinger writes a lengthy article entitled 'The present status of quantum mechanics', which is published in the November–December 1935 issue of the science magazine *Die Naturwissenschaften*. In it, Schrödinger summarises his understanding of the theory he helped to create, and coins a new term that will be part of every quantum scientist's vocabulary from now on: 'entanglement'. And he invents a cat:

> It is also possible to construct very burlesque cases. Imagine a cat locked up in a room of steel together with the following hellish machine (which has to be secured from direct attack by the cat): A tiny amount of radioactive material is placed inside a Geiger counter, *so* tiny that during one hour *perhaps* one of its atoms decays, but equally likely none. If it does decay, then the counter is triggered and activates, via a relay, a little hammer which breaks a container of prussic acid. After this system has been left alone for one hour, one can say that the cat is still alive *provided* no atom has decayed in the meantime. The first decay of an atom would have poisoned the cat. In terms of the ψ-function of the entire system this is expressed as a mixture of a living and a dead cat.
>
> Typical of these cases is that an uncertainty at the atomic level has been transformed into a coarse-grained uncertainty, which can then be *decided* by direct observation.

In other words: according to Niels Bohr and Werner Heisenberg's interpretation of quantum mechanics, the cat is both dead and alive—or neither dead nor alive. Such a thing is not possible, Einstein and Schrödinger agree. There is no such thing as a cat in a 'blurred' state between life and death. There is no such thing as a pile of gunpowder that has both exploded and not exploded. Quantum mechanics is not a description of reality.

However, Schrödinger is worried. He's concerned about the financial provisions for his and Anny's future. His German pension will not be enough to support them both, and they will not be able to live off the Schrödinger equation. 'It's not that I can't stand to stay in one place,' he writes to Einstein in May 1935, 'I've liked it everywhere I've lived so far, except in N … Germany. And it's not as if the people here weren't very kind and friendly to me. But that reinforces the feeling that I hold no position and have to rely on the generosity of others.' He wants a permanent post. But where? He had visited Princeton in 1934 and the university offered him a post, but it was too far away. He's tempted by an offer from the University of Madrid, but the outbreak of the Spanish Civil War puts an end to that.

Schrödinger receives an offer of a professorship in Edinburgh. He tells Hansi he will accept the post if she comes with him, but she refuses. So he moves to Graz, in his native Austria. Closer to Hilde and Ruth. But also closer to the Nazis.

The Nazis annex Austria in 1938. The university in Graz is closed down, and its rector and many of its teaching staff are sacked. Many Jewish members of staff are arrested; some leave all their belongings behind and flee abroad. Erwin Schrödinger can stay. He is not of Jewish descent, and tries to accommodate himself to the regime. But the Nazis know that he's not one of them, and they haven't forgotten his hasty departure from Berlin five years before. When the university reopens, the new rector presses Schrödinger to declare his support for the Nazis publicly. On 30 March 1938, shortly before a referendum on the German 'incorporation' of Austria, which had in fact already taken place, the *Grazer Tagespost* newspaper published a declaration by Schrödinger. It bore the title 'The Hand to Everyone Willing: a confession to the *Führer*—a top scientist reports for service for the people and the fatherland':

> In the midst of the exultant joy which is pervading our country, there also stand today those who indeed partake fully of this joy, but not without deep shame, because until the end they had not understood the right course. Thankfully we hear the true German word of peace: the hand to everyone willing, you wish to gladly clasp the generously outstretched hand while you pledge that you will be very happy, if in true cooperation and in accord with the will of the Führer you may be allowed to support the decision of his now united people with all your strength. It really goes without saying that for an old Austrian who loves his homeland, no other standpoint can come into question; that—to express it quite crudely—every 'no' in the ballot box is equivalent to national suicide. There ought no longer—we ask all to agree—to be as before in this land victors and vanquished, but a united people that puts forth its entire, undivided strength for the common goal of all Germans.
>
> Well-meaning friends, who overestimate the importance of my person, consider it right that the repentant confession that I made to them should be made public: I also belong to those who grasp the outstretched hand of peace, because, at my writing desk, I had misjudged up to the last the true will and the true destiny of my country. I make this confession willingly and joyfully. I believe it is spoken from the hearts of many, and I hope thereby to serve my homeland.

With this, Schrödinger provides the Nazis with a splendid piece of propaganda, astonishing all his friends abroad. He later regrets his confession of regret, apologises to Einstein for his 'cowardly writing', and explains his 'duplicity' by saying he just wanted the Nazis to leave him alone.

That doesn't work. When he returns from a summer holiday with Hilde in the Dolomite Mountains, he discovers his own job is being advertised. He is dismissed without notice, for 'political unreliability'. In his staff file, the university administration describes Schrödinger as 'excellent in his subject', but 'contradictory in his behaviour' and a 'semitophile'.

Schrödinger moves abroad again. He and Anny board a train to Rome with three hastily packed suitcases before their passports can be confiscated. They leave Schrödinger's Nobel Prize medal behind. They roam Europe, travelling to Geneva, then through France to England. Schrödinger is snubbed in Oxford, due to his public support of Hitler. Eventually, the University of Ghent in Belgium accepts him as a visiting professor, and Hilde March joins them there with her daughter, Ruth. Then the Second World War begins. Shortly before Germany invades Belgium, the Irish prime minister, Éamon de Valera, offers Schrödinger a permanent post at the newly established Institute for Advanced Studies in Dublin. Finally, he's found a place where he can stay with his family, consisting of two wives and a daughter. Schrödinger never owns a cat.

PRINCETON, 1935

Einstein puts the world back in focus

Princeton, 1935. Einstein has gained a measure of peace in his turbulent life. His great discoveries are now behind him. He's been the star scientist at the Institute for Advanced Study in Princeton for the past two years. It was the main objective in founding the institute to provide him with an academic home, and it will remain the last stop on his career journey. 'Princeton is a wonderful spot, and at the same time a quaint and ceremonious little village of tiny demigods on stilts,' he writes in a letter to Belgium's Queen Elisabeth.

However, he can't let go of the little-loved theory of quantum mechanics, which he is simply unable to accept. He can't just let this 'witch's multiplication table' go unchallenged, and remains steadfast in his belief that the real world exists independently of whether, and how, humans perceive it, while its secrets are nonetheless accessible to human understanding.

Over and over, Einstein attempts to uncover the weak points in quantum mechanics using his favourite tool, the thought experiment. He writes long letters to Schrödinger. He fills his chalkboard with tables and equations. In his office at 209 Fine Hall, he plays in his imagination with light particles, screens, boxes, scales, piles of gunpowder, and cats. He tries to forge his imaginings into sharp arguments, but he lacks support from others. His assistant, Walther Mayer, nicknamed 'Einstein's calculator', soon baulks at following his boss into battle against quantum mechanics.

Over afternoon tea one day in early 1934, Albert Einstein strikes up a conversation with the physicist Nathan Rosen from Brooklyn. At twenty-five, Rosen looks young for his age, has just married his high-school sweetheart, and is eager to prove himself as a scientist. The two physicists become engrossed in a conversation about quantum mechanics. Soon, they're joined by another of the institute's scientists, the thirty-seven-year-old Russian Boris Podolsky, who has previously worked with Paul Dirac.

Einstein wins Podolsky and Rosen over to his side as allies in his attack on quantum mechanics. The three scientists are soon referred to by the initials of their surnames, 'EPR'. They devise one of Einstein's thought experiments dealing with perhaps the oddest of quantum mechanics' oddities—'entanglement', as Erwin Schrödinger calls it. Entanglement does not fit with the traditional worldview of physics. Two entangled objects are somehow connected, despite being separated in space, as if they could speak to each other telepathically across thousands of kilometres or even light years.

Imagine a pair of light particles that are created together but then separate. For example, an 'excited' atom—one that has been shifted to a higher energy level—might radiate its superfluous energy as two light particles zooming off in different directions to the opposite edges of the solar system and beyond. According to quantum mechanics, the two particles are entangled: their properties are linked, no matter how far apart they are. This can be imagined in terms of colours—if one particle is red, the other is also red. If one is blue, then so is the other.

'What's so odd about that?' you might ask. It seems to be no stranger than a pair of socks, if one were sent to London and the other to Moscow. Their properties are linked, too: anyone who sees that the sock in London is black will immediately know that the sock in Moscow is also black.

But unlike the socks, the properties of quantum objects only become unambiguous when they are observed. The two light particles are in a 'superposed state' of being 'both red' and 'both blue'. It's only when one of the particles is observed that it becomes a certain colour;

and simultaneously, as if by magic, the other, far-distant particle takes on the same colour.

Now, that really is strange. How can the two particles take on the same colour? In their Copenhagen interpretation of quantum mechanics, Niels Bohr and Werner Heisenberg claim that neither of the two light particles is unambiguously red or blue until a measurement is made. Which particle is measured is a matter of chance. But if that is true, how can a random colour in London be the same in Moscow? If you toss one coin in London and another in Moscow, the result in one city doesn't influence the outcome in the other.

How can that be possible? Perhaps a very, very fast signal carries the colour between the two distant particles as soon as it is established? This concept is what Einstein dubs 'spooky action at a distance', and he considers it about as credible as a ghost story. It would be a contradiction of the fundamental principle at the base of his theory of relativity, according to which no effect can propagate faster than light—the principle of locality. There is something rotten in Copenhagen.

There remains only one possible explanation for the choreographed behaviour of the particles: their colour is fixed all along, like that of the socks. That's Einstein's belief, and he convinces Podolsky and Rosen of it. The three Princeton physicists take this as proof that quantum mechanics is an incomplete theory. It doesn't explain the underlying reason for the two particles having the same colour. It doesn't provide a sharper image of a blurred reality, as Bohr and Heisenberg claim. Rather, it provides a blurred image of a sharply focused reality.

The subsequent paper is authored by Podolsky—'for linguistic reasons', as Einstein explains in a letter to Schrödinger. Einstein considers his English not to be good enough. However, Podolsky's is not much better. 'Can quantum-mechanical description of physical reality be considered complete?' is the title of the article. There really should be a 'the' or an 'a' after the word 'can'—but Podolsky's native language is Russian, which has neither definite nor indefinite articles.

The paper is only four pages long, and closes with the lines, 'While we have thus shown that the wave function does not provide a complete description of the physical reality, we left open the question of whether or not such a description exists. We believe, however, that such a theory is possible.'

There is no time for corrections. Podolsky submits the manuscript in a hurry before Einstein has seen it, and disappears off to California. On 4 May 1935, eleven days before the article appears in the *Physical Review*, Einstein spots his name in a headline in the Saturday edition of the *New York Times*: 'Einstein Attacks Quantum Theory—Scientist and Two Colleagues Find It Is Not "Complete" Even Though "Correct"'. There follows a single-column summary of the article, with quotes from Boris Podolsky. The most famous scientist in the world is shaking the foundations of a new, six-time Nobel Prize–winning theory. This is the stuff of scandals—even if almost no one understands what it's actually about.

Einstein is angry. Podolsky is circulating the results of their research before they are published. On 7 May, the *New York Times* prints a statement by Einstein in which he declares that Podolsky has disclosed the results 'without my authority. It is my invariable practice to discuss scientific matters only in the appropriate forum and I deprecate advance publication of any announcement in regard to such matters in the secular press.' Einstein will never speak to Podolsky again following this incident.

Boris Podolsky's tittle-tattling to the press is not Albert Einstein's only reason for being angry with him. The Russian took a certain liberty in the article that Einstein would never have agreed to if he'd spotted it beforehand.

After arguing that quantum mechanics is incomplete, Podolsky goes further and attempts to refute Heisenberg's uncertainty principle with a rather dubious argument. That's a step too far, which leaves the article open to attack. It obscures Einstein's point. At first reading, the EPR paper looks like a clumsy attempt to outsmart the uncertainty principle. But Einstein doesn't care about the uncertainty principle. He has long since given up trying to disprove quantum mechanics, although there is nothing he would like better. 'What I wanted did not come out so well,' a disappointed Einstein writes to Schrödinger, 'rather, the essential thing was, so to speak, smothered by the formalism.'

In this, of all papers! It's Albert Einstein's final publication of lasting importance, and the most quoted one. The EPR paper causes a great stir among his colleagues on both sides of the Atlantic. 'Now we have to do it all over again. It doesn't work. Einstein proved it wrong,' complains Paul Dirac. Wolfgang Pauli calls the paper 'a catastrophe', and urges Werner Heisenberg to write a response in the *Physical Review* 'to prevent a certain risk of confusing public opinion — namely in America'. However, after learning that Niels Bohr is already working on a response, Heisenberg soon abandons his draft, preferring to leave the master himself to deal with this heresy.

In Copenhagen, the EPR paper strikes 'like a bolt from the blue', in the words of Bohr's colleague, Léon Rosenfeld. 'We dropped everything. We had to clear up such a misunderstanding immediately.' Bohr and Rosenfeld wrestle every day with a compelling response. Bohr struggles to understand the arguments coming from Princeton. 'What could they mean?' he asks Rosenfeld. 'Do you understand it?' With Rosenfeld's help, Bohr manages to produce an answer and to send it to the publishers of the *Physical Review* within six weeks, which is an extremely rapid response by his standards.

Bohr thoroughly dissects the EPR thought experiment in his own inimitable way. Not with equations, but with metaphors. No, he states, 'there is no question of a mechanical disturbance' between the far-distant particles. However, there remains 'the question of an influence on the very conditions that define the possible types of predictions regarding the future behaviour of the system'. Thus, Bohr distinguishes between 'influence' and 'mechanical disturbance'. What does he mean by this? That 'influence' can work instantaneously over vast distances, but 'disturbance' cannot? Perhaps. That quantum mechanics contradicts the principle of locality, which lies at the foundation of Einstein's theory of relativity? Maybe. No one understands it fully. Not even Bohr himself.

He will later admit that even he struggles to understand his own arguments, and his response article is another masterpiece of Bohr's typical impenetrability, full of convoluted sentences and obscure analogies. Although, years later, he will apologise for the 'inefficiency of expression' in his response to the EPR paradox, he never makes any attempt to redress it.

Very few physicists bother to burrow through Bohr's jumble of words. Many have simply had enough of old-fashioned philosophical arguments. The mere fact that Bohr has responded to the EPR paper removes any doubts they may have had about this.

Only Schrödinger agrees with Einstein. On 7 June, he writes in a letter from Oxford to Princeton, 'I am very pleased that in the work that just appeared in *P.R.* you have publicly called the dogmatic quantum mechanics to account over those things that we used to discuss so much in Berlin.' If only other physicists would realise it. Einstein turns out to be justified in his fear that the core of his criticism of quantum mechanics will be lost in Podolsky's presentation. He receives a great many letters from physicists defending Bohr's interpretation of quantum mechanics and claiming to have found mistakes in the EPR paper. But they don't agree on what those mistakes are. Most fail to identify the central dichotomy of quantum mechanics—that it must either be non-local or incomplete.

In another letter to Einstein, Schrödinger vents his annoyance at the 'not very bright' reactions to the EPR paper. 'It is as if one person said, "It is bitter cold in Chicago," and another answered, "That is a fallacy, it is very hot in Florida."'

One physicist whose reaction Schrödinger considers 'not very bright' is Max Born. He is 'extremely disappointed' by the Einstein–Podolsky–Rosen paper, which he had heard 'sensational' things about. On the other hand, Born also dislikes the fact that Bohr 'is often nebulous and obscure in expression'.

All this philosophical squabbling has become wearisome to Born and others. What's the point of it, anyway? Quantum mechanics works, after all. Many physicists of the younger generation just want to get on and use it, to apply it to the world, not to brood over it. Philosophising is something best left to the philosophers. There is research to be getting on with. Less than four years after the publication of the EPR paper, a world war breaks out, and their calculations will play a gigantic role in its outcome.

GARMISCH, 1936

Dirty snow

In 1936, Germany hosts the Olympic Games twice—in Garmisch-Partenkirchen in winter, and in Berlin in summer. The winter games are a dress rehearsal for the summer. Germany is under international scrutiny. A boycott movement forms in the US, Britain, France, and the Netherlands in protest at the Nazis' brutal, racist policies. Germany's sports officials know that the summer games in Berlin will be on the line if the winter games in Garmisch don't go off perfectly. The Nazis see the games as an opportunity to show the world how peace-loving, and how superior, they are. They forbid anti-Semitic attacks in Garmisch, and remove all the 'Jews not welcome' signs. The propaganda machine controls the organisation of the games down to the last detail, from the opening ceremony to the athletes' meal plans. Many visitors are taken in by the show and believe it—all too willingly. Oh, the Nazis are sometimes a little overzealous, it's true, but on closer inspection, they're not that bad.

At the opening ceremony on 6 February 1936, the parade of athletes marches through heavy snowfall into Garmisch-Partenkirchen's newly built Olympic stadium, watched by 60,000 spectators. Spotlights illuminate the scene from the surrounding mountains as the Olympic Flame is lit for the first time at a winter games. Wehrmacht gunners fire a salute. The Olympic Fanfare rings out, and Hitler declares the games open.

Adolf Hitler demonstratively leaves the stadium when the German ice hockey team plays. The team includes Rudi Ball—a 'half-Jew', according to the Nuremberg Race Laws passed just five months earlier.

Half a million spectators travel to Garmisch. A 30,000-strong crowd cheers as German skier Cristl Cranz, with her calm, rhythmical style, sets the fastest times in both slalom races, winning her gold in the women's combined Alpine event, even though she falls in the downhill race. Many of the spectators are keen partakers in the newly fashionable sport of downhill skiing themselves. The streets of Garmisch resound with the clatter of their steel-plated ski boots, and the buses and trains are packed with skiers, their rucksacks and skis on their backs.

To his own surprise, the American journalist and foreign correspondent William Shirer enjoys his stay in Garmisch-Partenkirchen, 'the scenery of the Bavarian Alps, particularly at sunrise and sunset, superb, the mountain air exhilarating, the rosy-cheeked girls in their skiing outfits generally attractive, the games exciting, especially the bone-breaking ski-jumping, the bob races, the hockey matches, and Sonja Henie'. The Norwegian figure skater Sonja Henie greets the Reich chancellor with such an enthusiastic Hitler salute that Norway's newspapers ask 'Is Sonja a Nazi?' Henie wins gold, and flirts with the Nazi bigwigs. After the games, Hitler invites her to lunch at his mountain retreat in Obersalzberg, where he gives her a photograph of himself with a dedication.

Three weeks after the end of the Olympic Games, in March 1926, German troops march into the demilitarised Rhineland. With this act, Adolf Hitler violates the Locarno Treaties, which paved the way for Germany to rejoin the international community after the First World War, and which helped end the isolation of German physicists.

The Hitler regime has Germany firmly in its grip, and no one can escape its influence. Including Werner Heisenberg. He would like to succeed his teacher, Arnold Sommerfeld, as professor of physics in Munich. Sommerfeld was sent into retirement in spring 1935, and Heisenberg is his preferred successor. But Heisenberg's nomination is opposed by Johannes Stark, the powerful president of both the Imperial Physical Technical Institute and the Association of German Science. Stark denounces Heisenberg as being not even the spirit of Einstein, but 'the spirit of the spirit of Einstein', and launches a campaign against him and the whole of theoretical physics. Back in 1926, he wrote in the

Nationalsozialistische Monatshefte (*National Socialist Monthly)*, 'Upon the sensation and advertisement of Einstein's relativity then followed the matrix theory of Heisenberg and the so-called wave mechanics of Schrödinger, each one as obscure and formalistic as the other.'

Werner Heisenberg gains a reputation for not being sufficiently anti-Jewish, and is accused of being a 'white Jew' and pursuing the 'un-German' science of theoretical physics. His attempts to defend himself are interpreted by some outside Germany as a sign that he is a willing supporter of the Nazi regime and a silent tolerator of the persecution of Jews. Max Born will later describe him as 'Nazified'.

Johannes Stark publishes an article in the SS newspaper *Das Schwarze Korps* (*The Black Corps*) in July 1937 under the title 'White Jews in science', which is primarily aimed at Werner Heisenberg. Stark calls Heisenberg a 'proconsul of Einstein's spirit in the new Germany' and the 'Ossietzky of physics', and he bemoans the fact that, following the 'elimination' of Jewish scientists, they still have 'defenders and successors among Aryan Jew-comrades and Jew-pupils'.

Carl von Ossietzky, a journalist, pacifist, and highly articulate critic of the Nazis, had been imprisoned since the Nazis seized power. In November 1936, he wins the Nobel Peace Prize for 1935, but is prevented from travelling to the award ceremony in Oslo by the Gestapo. Ossietzky dies in May 1938 as a result of the torture he suffered in several concentration camps, and the tuberculosis he contracted there.

Heisenberg sees his career is in jeopardy. Fearing he will be sidelined by the German scientific community, he does all he can to shed the comparison with Ossietzky and the reputation of being a 'white Jew'. He turns to Heinrich Himmler, Reichsführer of the SS, for protection. The Himmler and Heisenberg families have been acquainted since Himmler's father and Heisenberg's grandfather were in the same hiking club. Himmler leaves Heisenberg waiting for a response for a year, and Heisenberg starts planning his emigration. Then Himmler exonerates Heisenberg and condemns the 'offensive' against him in *Das Schwarze Korps*. 'I have had your case examined with particular care and scrutiny, since you were recommended to me by my family,' writes Himmler to Heisenberg. 'I am happy to be able to inform you today that I do not approve of the offensive against you in *Das Schwarze Korps* and that I have put a stop to any further attack on you.' He adds in a postscript, 'P.S. I believe, however, it is important that you make the distinction clear in front of your students, between the results of professional physics research and the personal and political attitudes of the scientists involved.' Heisenberg follows his advice and distances himself from Einstein: 'I have always followed Himmler's advice even before he gave it to me because I was never sympathetic towards Einstein's public conduct.'

However, even Himmler can't completely avert the damage to Heisenberg's career. The coveted professorship in Munich does not go to Heisenberg, but to a second-rate physicist called Wilhelm Müller, who is a member of the Nazi Party, a proponent of 'German physics', and, according to Sommerfeld, 'the worst successor imaginable'.

So Heisenberg has to stay in Leipzig, whether he likes it or not. The political situation affects his body and soul. So many of his colleagues have fled abroad, he seeks refuge in his private life.

On 28 January 1937, he organises a house concert at the home of the Bücking publishing family. Heisenberg, the host, and a violinist perform Beethoven's piano trio Op. 1 No. 2. In the audience sits twenty-two-year-old Elisabeth Schumacher, who has just abandoned her art history degree in Freiburg and moved to Leipzig to begin an apprenticeship as a bookseller. By the second movement of the piece, *Largo con espressione*, Elisabeth's heart has melted, and just two weeks later, Heisenberg writes to his mother, 'I got engaged yesterday—provided you consent.' Their wedding takes place in April. Eight months later, Elisabeth gives birth to twins, Wolfgang and Maria.

The next year, Heisenberg comes across an uninhabited timber house in the lakeside Alpine village of Urfeld am Walchensee. He buys it from Wilhelmine Corinth, daughter of the painter Lovis Corinth, for 26,000 marks as a refuge for his family, should war break out. It will be only a year before that happens and the next world war begins. Werner Heisenberg receives a conscription order from Berlin. In Nazi Germany, nothing private is safe from the clutches of the state.

MOSCOW, 1937

On the other side

Princeton, 1935. Paul Dirac is at the height of his career. His office at the Institute for Advanced Study is on the same corridor as Albert Einstein's, just two doors down. He's currently putting the finishing touches to the second edition of his masterpiece, *The Principles of Quantum Mechanics*, which some colleagues call 'the bible of modern physics'. Dirac's father, with whom he always had such a difficult relationship, dies in 1936. Dirac does not mourn him. 'I feel much freer now,' he writes to Margit Wigner, sister of the Hungarian physicist Eugene Wigner. Margit is candid, forthcoming, and talkative, and she has taken a liking to the shy Dirac. Paul affectionately calls her Manci. The couple marry in 1937, and Paul officially adopts Manci's two children from her first marriage, Gabriel and Judy. 'You have made a wonderful alteration in my life,' Dirac tells his wife after they are married. 'You have made me human.'

However, there is something that constantly preys on Dirac's mind. Shortly before Christmas 1934, he received a letter from Cambridge, sent by his best friend's wife, Anna Kapitza. It was a plea for help. Peter Kapitza had been stuck in the Soviet Union for months. 'I am writing to you as a friend of K. and of Russia, and you will understand the impossible situation,' Anna tells Dirac, who is immediately alarmed.

The background story is that Peter Kapitza had moved to England in 1921 to study experimental physics, leaving a Russia that was ravaged by war and revolution. One colleague described him as seeming like 'a sad prince'. He had lost four of his closest relatives within the space of four months in 1919. His son died of scarlet fever. Then his daughter, his wife, and his father all succumbed to the Spanish flu. Kapitza also caught the flu himself, but survived it. During the war he served as an ambulance driver for two years.

The widowed Kapitza had moved to Cambridge with a very clear idea of his aim — to work with Ernest Rutherford at the Cavendish Laboratory. Rutherford initially rejects him, but Kapitza is persistent, and Rutherford eventually takes him on. Kapitza develops a technique for creating ultra-strong magnetic fields by injecting high current for brief periods into specially constructed hollow electromagnets.

He admires and idolises Rutherford. Accepting this adulation, Rutherford grows far closer to Kapitza than is usual in a teacher–student relationship, and more than the social rules of the day allow. Kapitza becomes the son that Rutherford never had.

When Rutherford is out of earshot, Kapitza calls him 'the crocodile'. It's the greatest compliment he could pay him, as the crocodile is his favourite animal. He collects poems about crocodiles. He has welded a metal crocodile to the radiator grille of his car. Even physicists, or perhaps especially physicists, have their personal quirks.

When Paul Dirac arrives in Cambridge to work at the Cavendish Laboratory in 1921, shortly after Kapitza's arrival, the 'sad prince' has transformed into the most prominent, popular, and vibrant person in town. The well-built Russian doesn't really speak any language well, not English or French, or even his native Russian. Mixing them all together, he speaks 'Kapitzian', and he speaks it a lot. Bubbling over with words, he recounts stories, explains card tricks. Kapitza is the polar opposite of the taciturn Dirac. Kapitza loves chatter; Dirac prefers silence. Kapitza loves the theatre; Dirac considers it a waste of time. But they share a love of knowledge and an interest in the fundamental principles of the material world. Dirac is impressed by Kapitza's courage and daring. The two men become best friends.

Kapitza is an avid communist. He's so keen, in fact, that Rutherford has to ban him from spreading communist propaganda on his first day at the Cavendish Laboratory. Kapitza never joins the Communist Party, but he's committed to the revolutionary cause, and returns home to the Soviet Union every year to help with Stalin's industrialisation program. He is quoted as saying, 'I am in complete sympathy with the socialist reconstruction directed by the working class and with the broad internationalism of the Soviet government under the guidance of the Communist Party.'

It's a dangerous gamble, as his return to Cambridge is never guaranteed, and it becomes increasingly risky every year. Fear of communism is rampant in 1920s Britain. Kapitza's reputation as a 'bolshy' spreads quickly. The secret service, MI5, has its sights on him, and has him shadowed by a special police unit.

When the economist John Maynard Keynes hears in October 1925 that Kapitza is once again heading to the Soviet Union to assist the Council of People's Commissars with their plan for the electrification of the country, he warns, 'I believe they will catch him sooner or later.' Oh, nonsense, is Kapitza's reaction. He trusts the state's promises that he will always be allowed to travel abroad.

Kapitza also starts something of a revolution in Cambridge. Unhappy with the submissive attitude of young physicists towards their professors, he founds the Kapitza Club in October 1922 for the open and informal discussion of physics. The Kapitza Club meets every Tuesday evening in one of the rooms at Trinity College — attendance is by invitation only. They talk about their research without a manuscript, just with some chalk and a blackboard resting on an easel. Questions are allowed at any time. A fire blazes in the hearth, and most attendees sit on the floor due to a lack of sufficient chairs. Sometimes, a vote is held on whether the speaker is right or not. Werner Heisenberg speaks at the Kapitza Club in 1925; Erwin Schrödinger does the same in 1928.

In September 1934, Kapitza's passport is confiscated while he, along with his family, is visiting his mother in Russia. His wife, Anna, is allowed to travel back to Cambridge with their sons, six-year-old Sergey and three-year-old Andrey, while Kapitza himself is held in a miserably furnished hotel room in Moscow. Three months go by. Anna is desperate. She writes to Dirac, telling him that being detained in the Soviet Union is a 'terrible blow' for her husband after all he's done for the country. In fact, it is 'probably the most terrible blow of his life'. She fears that the press will get hold of the story and that 'people will talk'—destroying her chances of ever seeing her husband again. She begs Dirac, who won a Nobel Prize in the same year as Kapitza, to put a good word in for her husband with the Russian ambassador in Washington. 'I think that's the only way we can do anything at all.'

Dirac launches a campaign to gain Kapitza's release. He asks the advice of Albert Einstein and Abraham Flexner, the director of the Institute for Advanced Study, and also urges the crocodile, Ernest Rutherford, to take up Kapitza's cause. Rutherford pulls all the diplomatic strings he can, but to no avail. 'I feel like a virgin who was raped when she believed she was giving herself up for love,' Kapitza laments in a letter sent from the Soviet Union. Just by writing such lines, he risks deportation to the labour camps. The Soviet censors read his letters before they're sent, and the British secret service reads them after they arrive.

Dirac takes on the role of father to Sergey and Andrey in Cambridge. He takes the Kapitza boys on day trips in his clapped-out old car, and on 5 November puts on a fireworks display for them on Guy Fawkes Night, the annual commemoration of the 1605 Gunpowder Plot. He puts his work on pause for months to take care of them. When all else fails, he and his new wife, Manci, head to the Soviet Union in summer 1937, against the advice of their friends.

For many people in the Soviet Union, 1937 is the year of the worst horrors so far: the culmination of the 'Great Purge', with arrests, torture, and murder. Stalin's campaign to intimidate the population is brutal and chaotic. By the end of 1937, the Great Purge has claimed some four million lives. Neither Dirac nor Kapitza is aware of the extent of this fear and horror. Socialism remains their political ideal.

In July 1937, just days before Joseph Stalin legalises the torture of suspected enemies of the people, the Diracs arrive at Kapitza's country dacha in Bolshevo, north-east of Moscow. The weather is oppressively hot.

Paul and Manci stay three weeks. In the cool of the morning, Dirac and Kapitza fell trees and gather strawberries. In the shade on the veranda, Paul tells tales of the crocodile, Rutherford. Manci struggles with the sparse conditions at the dacha, where not even toilet paper is available. None of the party suspects that people are being tortured, killed, and buried in the woods nearby. Then it's time for the Diracs to leave.

Peter Kapitza is forced to stay in Moscow. The state builds an institute especially for him, where he continues the experiments he had to leave behind in Cambridge. Soon after the Diracs' visit, Peter discovers 'superfluid' helium, which has practically no viscosity at very low temperatures and appears to defy gravity by flowing up walls. This strange behaviour can only be explained by quantum mechanics. Kapitza becomes the father of low-temperature physics.

Dirac and Kapitza sense that they will not see each other again for a long time. It is not until twenty-nine years later, in 1966, that they meet again, in the nuclear age, long after the start of the Cold War.

BERLIN, 1938

Bursting nuclei

Berlin, 1907. The twenty-eight-year-old chemist Otto Hahn is an oddity at the university's chemical institute. The subject of his research is radioactivity. 'Is that even chemistry?' ask his long-established colleagues, 'or perhaps alchemy?' Hahn specialises in creating new elements in the lab. Didn't the trickster and alchemist Cagliostro try to do precisely that back in the eighteenth century? Hahn is now sitting in the office of Emil Fischer, director of the institute and winner of a Nobel Prize. He's asking for something unheard of. He wants Lise Meitner to come and work in his laboratory. A woman, and a physicist to boot: she's not even a chemist. Fischer doesn't allow women to attend his lectures, let alone become researchers at his institute, where the only women allowed are the cleaners. 'I'm certainly not going to start running a henhouse here!' Fischer responds.

Hahn eventually manages to persuade Fischer to let Meitner work with him—albeit under some very strict conditions. Hahn and Meitner's laboratory is in a former woodworking shop in the basement of the institute. Frau Meitner must not enter the main institute without permission. The lab has a separate entrance, which Meitner must always use. There is no ladies' toilet, so Meitner has to use the facilities at a nearby inn. Sometimes she hides under a bench in the auditorium to listen to lectures. She works as an 'unpaid guest', receiving no salary and relying on money sent to her by her parents in Vienna. Lise Meitner lives off bread and black coffee.

She has been living with injustices such as these for three decades. She was born in Vienna in 1878, at a time when women had no access to higher education. Young women's only function was to give birth to babies. In the exceptional case that they did work, at most it was only as school mistresses. Lise, who had been interested in science since her school days, is barred from attending any institute of higher education. After eight years at a girls' school, she starts training as a French teacher, while simultaneously preparing on her own for the external school-graduation certificate, which is required to enter university-level studies. She passes it at the age of twenty-two. The University of Vienna has now started accepting female students, at least. Lise Meitner ups the tempo. She enrols in university courses in mathematics, philosophy, and physics; studies under Ludwig Boltzmann; and gains her doctorate after just eight semesters — only the second woman to gain that honour from the University of Vienna. At the age of twenty-eight, she heads to Berlin, the capital of physics, where she will stay for the next thirty years. She wants to learn more and study under Max Planck. But women are still not allowed to study in Berlin. Never daunted, Lise Meitner dares to approach the 'Privy Councillor' Planck, telling him she has come to Berlin to gain 'a true understanding of physics'. Astonished, he responds to this slightly built, initially shy-looking woman, 'You have your doctorate, what more do you want?' Planck is a traditionalist and has spoken out against 'female Amazons' in science in the past. But he listens to what she has to say, and realises Lise Meitner deserves to have an exception made for her. He allows her to enrol for his lectures.

Lise Meitner doesn't complain about the discrimination she faces. She lets her actions be her response to the deep-seated prejudice against women that most men harbour. In their first year of experimenting together, Hahn and Meitner produce a number of new isotopes: new atoms of previously known elements with a different mass. News of Meitner's talent as a scientist soon spreads. In 1909, she delivers a lecture on beta radiation at a congress in Vienna, and a year later meets the Nobel Prize laureate Marie Curie in Brussels. She will later describe the years she spent working in the cellar at the institute as the happiest of her life. 'We were young, merry and carefree; politically speaking, perhaps too carefree.'

After a couple of years, Max Planck makes her the first female university assistant in Prussia. For Lise Meitner, the job is 'like a passport' to the impenetrably male-dominated scientific community and 'a great help in overcoming the existing prejudice against female scholars'. Albert Einstein calls her 'the German Marie Curie'.

The papers she publishes in the *Naturwissenschaftliche Rundschau*, the *Scientific Review*, are signed only with her surname, leading many readers to assume they are written by a man. The encyclopaedia publisher Brockhaus sends a letter to the supposed 'Herr Meitner' in an attempt to get 'him' to write an entry in one of its encyclopaedias. When Lise Meitner outs herself as a woman, the publisher immediately loses interest.

Perhaps Lise Meitner might have stagnated as an 'unpaid guest' in Otto Hahn's research group if the University of Prague had not offered her a post as a lecturer. When that happens, the Prussians suddenly realise the asset they have in Meitner. In 1913, at the age of thirty-five, she is finally offered a permanent position at the Kaiser Wilhelm Institute for Chemistry. She's delighted by 'the wonderfulness of science', and can finally pay for her own coffee.

When the First World War breaks out, there are suddenly more important things than science. Lise Meitner volunteers as an X-ray nurse-technician, and serves on the Austrian eastern front. Deeply affected by the horrors of war, she returns from the front in 1916 and insists that she be allowed to continue to work with Otto Hahn in their old laboratory—if he is in Berlin. Hahn is serving in the 'gas regiment', developing poisoned gas in the 'special unit for chemical warfare' run by Fritz Haber. Hahn trains soldiers to prepare them for poison gas attacks, and commutes between the eastern, western, and southern fronts, and research facilities in Berlin and Leverkusen. It's only when he's on home leave that he has time to join Lise Meitner in the laboratory. Thus, it's mainly down to Meitner that the two of them manage to produce samples of the new element protactinium in 1917, shortly before the end of the war. The new element is stable, radioactive, and has the atomic number 91.

The end of the rule of the House of Hohenzollern and the birth of the republic on 9 November 1918 bring some degree of liberty for women, including Lise Meitner. From 1920, women are able to gain the post-doctoral professorial qualification known as the habilitation. Meitner is awarded her habilitation immediately, on the strength of her previous achievements—thirteen years later than Otto Hahn, who is the same age as her. Meitner's inaugural lecture is on 'The Significance of Radioactivity for Cosmic Processes'. Typically, one journalist turns the title into 'Cosmetic Processes'. Meitner can only scoff at such lazy male assumptions about gender roles. She overtakes one male competitor after another, and is soon running her own department at the Kaiser Wilhelm Institute, researching beta decay and gamma radiation, and attending conferences all round the world. She moves into a large apartment in the Kaiser Wilhelm Institute director's villa in the leafy academic suburb of Dahlem. In 1926, she is named extraordinary professor of nuclear physics, making her the first female physics professor in Germany.

Hahn and Meitner make a strange couple—and their relationship remains purely scientific. They spend thirty years working very closely together, encouraging and supporting each other. But, except for official occasions, they never eat together, never go out for a walk together, and never visit each other at home. They remain on formal, last-name terms until well into the 1920s. Hahn gets married in 1913.

Meitner never marries.

She doesn't miss marriage or children. 'I simply never had the time for that,' she says. Her co-workers are her family. Feminism? She doesn't need it. Later, she realises 'how wrong my opinion was and how much gratitude any woman working in academia owes to the women who fought for equality'.

Science is her life, and her life is going well. But that comes to an end in 1933, when the Nazis seize power. Many of Meitner's colleagues who are of Jewish descent leave the country. Lise Meitner was baptised in the Lutheran church and received a liberal education, but in the birth register she is recorded as being of Jewish descent. For the Nazis, that makes her 'non-Aryan'. Now aged fifty-five, she refuses to abandon her life's work and leave Berlin. She turns down job offers from abroad, preferring to stick it out in the German capital. That's a decision she will later regret. 'Today, I know,' she will say after the war, 'that it was not only stupid, but also grossly unjust, because by staying, I ultimately gave my support to Hitlerism.'

At least Meitner is protected from even worse discrimination by her Austrian citizenship. She continues her research, and persuades Hahn to return to their collaboration. Their aim is to produce atoms that are heavier than the heaviest known element, uranium: transuranic elements. They want to fill in the gaps in the periodic table beyond uranium.

In an attempt to achieve this, Meitner and Hahn fire neutrons, which have recently been discovered at the Cavendish Laboratory in Cambridge, at uranium atoms, hoping the neutrons will penetrate the atoms' nuclei. In 1935, Lise Meitner, Otto Hahn, and a young chemist called Fritz Strassmann begin a series of experiments, and no one dreams how literally explosive their results will be. These experiments will change the course of world history.

But before this happens, world history changes the course of their experiments. Lise Meitner's Austrian passport becomes invalid with the 'annexation' of Austria by the German Reich. From one day to the next, she now faces persecution as a 'Jew of the German Reich', with no official protection. She is no longer allowed to work, but is also forbidden from going abroad. The chemist Kurt Hess, a Nazi Party member and Meitner's adversary, denounces her for 'endangering the institute'. She tries to persuade the interior ministry to issue her with valid German papers. Without success.

Lise Meitner is trapped.

Her only option now is to flee. On 13 July 1938, with only ninety minutes' notice, she packs a few important possessions. Friends smuggle her across the border into the Netherlands, from where she travels on through Denmark to the safe haven of Stockholm, where the Nobel institute offers her a temporary job.

In Stockholm, Meitner might be physically removed from the nuclear research laboratory at the Kaiser Wilhelm Institute, but she is not cut off from the progress of research there. Hahn and Strassmann secretly keep her posted by letter. In that way, according to Strassmann, she 'remained the spiritual leader of our team'.

The events of autumn 1938 show that Lise Meitner did the right thing by fleeing Germany. The 'night of broken glass' on 9 November 1938 marks the start of Europe's worst-ever genocide. All over Germany, Jewish shops and synagogues are destroyed and plundered by the Nazis. They throw stones through shop windows, chop up furniture with axes. More than a hundred Jewish people are murdered, thousands more are arrested and sent to the concentration camps, and countless others suffer abuse, beatings, and public humiliation. The fire brigade stands by, making sure that the flames from the burning Jewish shops don't spread to other buildings. There's so much shattered glass that the Nazis dub the action 'Reich Crystal Night'. It was organised centrally from Berlin. Not wanting his name to be associated with this pogrom, Adolf Hitler leaves the 'organisational work' to Hermann Göring and Joseph Goebbels.

Hahn and Strassmann toil away, day after day, in their lab, but make little progress. For months now, they've been trying to produce elements that are heavier than uranium; but, strangely, their chemical analyses seem to show the presence of far lighter atoms. They appear to be atoms of radium, but that makes no sense. Where are these atoms coming from? At a loss to explain this, Otto Hahn travels to Copenhagen in November to consult with Niels Bohr, the master of the atom. Lise Meitner joins them in Copenhagen, along with her nephew, the physicist Otto Frisch. So many intelligent minds, but none of them can explain those results.

Returning to Berlin to continue their work, Hahn and Strassmann keep coming up with ever stranger results. They discover that the mystery atoms are not radium, but barium. With an atomic mass of just over 137, barium is only half as heavy as uranium-238. Can it be that the beam of neutrons fired at the uranium atoms is bursting them apart? Hahn simply does not have the physical imagination to come up with or express such a hypothesis. He's an excellent experimental scientist, but not an exception to the 'law of the conservation of genius': brilliant experimentalists make terrible theoreticians, and vice versa.

In a letter sent to Lise Meitner in Stockholm on 19 December 1938, Otto Hahn's confusion and desperation are plain: 'It's now just 11 pm; at 11.45 Strassman is coming back so that I can eventually go home. Actually, there is something about the "radium isotopes" that is so remarkable that for now we are telling only you … It could still be an extremely strange coincidence. But we are coming steadily closer to the frightful conclusion: our radium isotopes do not act like radium but like barium.'

What is happening to the uranium? 'We know ourselves that it *can't* actually burst apart into barium,' Hahn continues in his letter to Meitner. 'Perhaps you can come up with some fantastic explanation?' No, she can't. At least not that quickly; she needs time to do her calculations. 'The assumption of such a thorough-going bursting seems very difficult to me,' Meitner concurs in her reply of 21 December 1938, 'but in nuclear physics we have experienced so many surprises that one cannot unconditionally say: it is impossible.'

On the same day, Hahn writes to Meitner again, pushing for an explanation, and their letters cross in the mail. 'We cannot suppress our results, even if they are perhaps physically absurd. You see, you will be doing a good deed if you can find a way out of this.'

During a winter walk, Lise Meitner and Otto Frisch discuss the mysterious measurements from Berlin. Perching on a tree trunk in the snow-covered Swedish forest, they scribble their musings down on scraps of paper. They come up with a new model of the atomic nucleus. A heavy nucleus can become 'wobbly', like a drop of water, when hit by a neutron. If it undergoes sufficient deformation in the process, the long-range electrostatic repulsion is enough to overcome the short-range nuclear force holding it together. The nucleus shatters in an explosion. Meitner and Frisch are able to estimate the explosion's energy using Einstein's equation, $E=mc^2$. It is huge.

Returning to Copenhagen, Frisch tells Bohr about the theory. Slapping his forehead with his palm, Bohr declares, 'Oh, what idiots we have all been! We could have foreseen this.' However, Bohr now foresees something else that sends chills down his spine: the potential destructive power of this energy released from an atomic nucleus. And that potential destruction will become actual far sooner than any of the physicists imagine. It will leave a dark cloud hanging over the brilliant years of physics research.

THE ATLANTIC, 1939

Terrible news

In January 1939, Niels Bohr and his assistant, Léon Rosenfeld, are on a steamer crossing the Atlantic to New York in order to visit Albert Einstein in Princeton. They are bringing terrible news with them: Otto Hahn has split the atom. He and Fritz Strassmann have managed to 'burst' uranium nuclei by bombarding them with neutrons.

Only four days before the ocean liner carrying Bohr and Rosenfeld set sail, Otto Frisch had broken the news in Copenhagen, and Bohr immediately recognised its huge implications. In Nazi Germany, of all places, physicists have taken the first step towards the development of a nuclear bomb.

Bohr's original reason for travelling to America has been to continue his ten-year conversation with Albert Einstein about quantum mechanics. But the grandmaster of quantum mechanics feels his interest in the subject waning. During the crossing, his mind wanders further and further from quantum mechanics, towards nuclear physics.

He spends the spring semester in Princeton. Every day, from January to April 1939, he thinks and talks about nuclear fission, scribbling equations on chalkboards and erasing them again. His main discussion partner is the American John Wheeler, who had studied under him in Copenhagen and is now a lecturer at Princeton. Einstein stays out of it. He's not interested in the niceties of nuclear fission, and doubts it can ever be put to any use.

Working with Wheeler, Bohr begins to play out the splitting of uranium nuclei in his mind. His musings are based on the work of Lise Meitner and her nephew, Otto Frisch, who fled their homes to work in Scandinavia, where they had come up with the theory of nuclear fission. Would it be possible to trigger a cascade of nuclear fissions, in which the energy from one nucleus splitting would cause the next to split, releasing yet more energy and so on, in a chain reaction? Is such a nuclear bomb conceivable?

Again and again, Niels Bohr and John Wheeler run through the possible ways a chain reaction might be triggered, until they decide to share the problem with two colleagues from Hungary, Leó Szilárd and Eugene Wigner. They keep their meetings highly secret. What they're dealing with has already gone far beyond pure science.

On 15 March 1939, Bohr, Wheeler, Szilárd, and Wigner are holding a long meeting in an empty office—the one next to Wigner's—in Princeton University's Fine Hall. This was where Albert Einstein had pursued his lonely quest for a unified field theory, before moving into the newly built premises of the Institute for Advanced Study just a few weeks before. Bohr has had a major insight a couple of days earlier, during a stroll from the Princeton Campus Club to the Fine Hall, when he realised that the nuclear fission Otto Hahn had measured occurs only in the rare isotope uranium-235, but not in the much more common isotope uranium-238.

It's a paradox. The uranium nucleus needs to be bigger for it to become fissile. But it can't be too big, either. Additional neutrons can cause it both to burst and to stabilise. Bohr and Wheeler calculate that the two isotopes, uranium-235 and uranium-238, have very different properties. When a neutron collides with the nucleus of an atom of uranium-235, that nucleus splits into two lighter nuclei, releasing a huge amount of energy and a couple of neutrons, which go on to split more nuclei. If the amount of uranium-235 is large enough—if it reaches 'critical mass'—a collision cascade will ensue, leading to a chain reaction. A lump of uranium about the size of a bowling ball would be enough to destroy an entire city. Or to supply that city with electricity for days, if the chain reaction could be controlled.

The heavier isotope, uranium-238, behaves rather differently. The three extra neutrons stabilise the nucleus, which cannot therefore be split so easily by other neutrons. It is not light enough for a chain reaction to take hold. Ninety-nine per cent of natural, mined uranium is U-238. 'So let's separate the U-235,' suggests Szilárd, 'and build a bomb out of it.'

'That would be possible,' replies Bohr, 'but it would mean turning the entire United States into one big factory.' Natural uranium contains only 0.7 per cent U-235. Extracting it would require a huge technical effort in industrial plants containing rows of centrifuges in their hundreds.

But what's the alternative? If they don't do it, the American war strategists believe, the Nazis will. The German Wehrmacht occupied Bohemia in the autumn of 1938. This means the Nazis have access to the uranium mines of Johannisthal, which had supplied the Curies with pitchblende thirty years earlier. That's a reason for concern and a reason to make haste. Leading German researchers formed the 'Uranium Club' in the spring of 1939, with the aim of developing a 'uranium machine' (nuclear reactor). American physicists and the secret services fear that the Nazis have a head start on them when it comes to research into harnessing the power of nuclear fission.

In summer 1939, Werner Heisenberg travels to the US for a lecture tour. In Michigan, he visits Enrico Fermi, the physicist from Rome whose wife is Jewish and who fled the fascist race laws in his homeland. Also there is Samuel Goudsmit, the Dutch physicist who had discovered electron spin as a shy young student. After emigrating to America, Goudsmit had changed the spelling of his name from Goudschmidt to Goudsmit, to preserve the correct pronunciation.

'Why don't you stay here in the USA?' Fermi asks Heisenberg. 'Then you'd be on the side of decency and civilisation.'

'It would feel like committing treason,' answers Heisenberg, and adds a sentence that sticks in Goudsmit's mind: 'Germany needs me.'

'What if Hitler forces you to build a nuclear bomb?' Fermi asks.

Heisenberg believes the outcome of the war will be decided before such a bomb is ready.

One Sunday afternoon, while picnicking with Wheeler, Heisenberg announces that he has to return home soon for 'machine gun shooting practice in the Bavarian Alps'. Heisenberg sails back across the Atlantic on the almost-empty steamship *Europa*. He's called up for military service a few weeks later. To his great surprise, he's not conscripted into the Alpine division, but to the *Heereswaffenamt*—the military ordinance office.

The Hitler regime takes control of the Uranium Club after the start of the war. The military ordinance office is drawing up 'preparatory working plans for the commencement of experiments on the exploitation of nuclear fission'. Carl Friedrich von Weizsäcker transfers to Berlin to work on the project. Werner Heisenberg designs a uranium reactor with heavy water serving as the 'braking substance' for the neutrons ejected during nuclear fission. Only slow-moving neutrons can go on to split more atomic nuclei.

Writing in a report for the Uranium Club, Heisenberg says, 'The enrichment of uranium-235 still remains the only method of producing explosives several orders of magnitude more powerful than the strongest explosives yet known.' His idea of a bomb that could 'blow up New York in white heat' makes an impression on those who run the military ordinance office.

Now aged sixty, Albert Einstein yearns for some peace and quiet, and withdraws to his holiday home on Long Island in the summer. All he wants to do is to play his violin, sail on the Atlantic, and read in the shade of the chestnut trees. But even there, he can't escape the horrors of the world. Leó Szilárd pays him a visit that summer. The two men know each other from Berlin, where they developed an 'automatic concrete people's refrigerator' together in the 1920s. Now Szilárd urges Einstein to do something to counter the nuclear threat from Germany. Maybe he could write a letter to the Belgian government? The largest uranium deposits in the world are in the Belgian colony of the Congo. Perhaps they can still be prevented from falling into the hands of the Germans? No, they can't, both Einstein and Szilárd realise. Just a few months later, the Wehrmacht occupies Belgium, and soon begins delivering thousands of tonnes of uranium ore from the Belgian Congo to Berlin.

Einstein and Szilárd decide to write to the American president, Franklin Delano Roosevelt. Einstein dictates the letter in German, and

Szilárd produces a clean English version. The self-styled 'convinced pacifist' Einstein urges Roosevelt to promote the development of a nuclear bomb. He warns the president that the recent discovery of nuclear fission could lead to the construction of 'extremely powerful bombs of a new type', and proposes a plan to accelerate research into nuclear fission for military purposes. Einstein points out that such research is already under way at the Kaiser Wilhelm Institute in Berlin, naming 'the son of the German Under-Secretary of State, Carl Friedrich von Weizsäcker' as a key person involved.

Einstein signs the letter in Peconic, Long Island, on 2 August 1939, with the words, 'Yours very truly, Albert Einstein'. Two years later, he will describe signing the letter as 'the one great mistake in my life. Had I known that the Germans would not succeed in producing an atomic bomb, I would not have lifted a finger.'

Roosevelt doesn't have time for letters from nuclear scientists right now. The global situation is grim. Hitler has invaded Poland, and Britain and France have declared war on Germany. Einstein and Szilárd's letter doesn't reach Roosevelt's desk until 11 October 1939. Roosevelt decides something must be done 'to see that the Nazis don't blow us up'. On that same day, he launches the Manhattan Project with the aim of developing a nuclear bomb. Following a sluggish start, the project gains momentum after Einstein sends two more letters to Roosevelt with suggestions for how to organise the project and containing further warnings about Germany's bomb-makers. The Manhattan Project not only turns the United States into a uranium factory, but it also unites the scientific forces of three countries—Britain, Canada, and the United States.

Many of the world's greatest physicists are involved in the Manhattan Project, including some who have fled Germany or one of its allied states. Otto Frisch, now based in England, calculates that just 50 kilograms of U-235 would have the explosive power of 15,000 tonnes of TNT.

Bohr returns to Copenhagen from America in May 1939. On 1 September 1939, German troops cross the border into Poland. On the same day, the *Physical Review* publishes an article by Bohr and

Wheeler, with the title 'The mechanism of nuclear fission'. It contains not a single mention of a chain reaction.

Before dawn on 9 April of the next year, the German Wehrmacht invades Denmark. The Danish government capitulates within only two hours. Adolf Hitler plans to establish a 'model protectorate' in Denmark as a demonstration to the rest of the world of his peaceful intentions. Niels Bohr, the man who has given refuge to so many fleeing physicists in the past, will soon have to flee himself.

COPENHAGEN, 1941

Estrangement

Copenhagen, 16 September 1941, late evening. Two men are taking a stroll. They have often walked together here in Copenhagen since their first stroll twenty-two years earlier in Göttingen — they are Niels Bohr, now aged fifty-five, and Werner Heisenberg, now thirty-nine. Back then, Heisenberg was Bohr's protégé. Later, they became colleagues, jointly founding the discipline of quantum mechanics. That first walk in Göttingen now feels like an age ago. It was here, on a nocturnal meander through the Fælledparken gardens, that Heisenberg conceived of his uncertainty principle. Both men have grown older, and their pace has grown slower. They are now on opposing sides of a war. Father and son have become estranged.

Heisenberg has travelled from Berlin under the pretext of giving a lecture on cosmic rays. He's in Copenhagen for a full week, but he feels the need to speak with Bohr immediately; so, as soon as he arrives, he hurries to Bohr's house and greets Margrethe and the Bohrs' sons. The two scientists choose to take a walk in the park, where they can be sure no one is listening in on their conversation. Heisenberg is eager to get straight to the point: the bomb. He has placed so much hope in this conversation that his perception of the talk is coloured by it. There is no longer any of the old familiarity between them. Something now stands between Bohr and Heisenberg that was never there before, even subconsciously: mistrust.

Bohr is taken aback. The way Heisenberg speaks has changed; the Nazi jargon is obvious to Bohr. He's alarmed by Heisenberg's prophecy that physics will play a prominent role in a Europe under

Hitler's control. What does Heisenberg want? Why has he come to the occupied Danish capital? Is he a friend or a foe? Is he a Gestapo spy? Does he want to sound Bohr out? Or protect him?

Bohr's mother came from a Jewish family, making him a 'half Jew' according to the Nazis' race categories. He's also involved in the resistance to the Nazi occupation of his country, and has had contact with physicists involved in America's nuclear-bomb project. He is under surveillance by the German secret service. He's constantly on the verge of being arrested. Only a few weeks earlier, the police incarcerated 300 Danish communists in the Horserød camp on the island of Zealand.

Heisenberg is also heavy-hearted. He recently received news that his favourite student, Hans Euler from South Tyrol, is registered as 'missing on the Eastern Front' after his plane crashed. As a communist and opponent of the Nazis, Euler had refused to collaborate with the uranium project. Heisenberg supported and protected him. Following a personal crisis, Euler volunteered for military service, and worked as a meteorologist and navigator for the air force's weather reconnaissance unit. His colleague Carl Friedrich von Weizsäcker wrote that it was 'as if he basically sought death'. Shortly after Germany invaded the Soviet Union on 23 April 1941, the engine of Euler's plane was damaged by gunfire. After ditching into the Sea of Azov, the crew were taken captive by local fishermen. Heisenberg never finds out what became of Euler after that, despite his attempts to learn the fate of his favourite student.

No one in Europe has solid ground under their feet anymore. No one knows how this war will end. Having defeated France, the Germans have the upper hand and are now in control of much of the continent. The Wehrmacht is rapidly advancing on Moscow, and Heisenberg speaks of a German victory in that campaign as if it were a foregone conclusion. However, German troops have not yet faced the Battle of Stalingrad.

Werner Heisenberg is the uranium project's leading scientist. He's working on the construction of a 'uranium burner' to supply the war effort with energy, which in miniaturised form could also power

German tanks and U-boats. He and von Weizsäcker have discovered another element, besides uranium-235. Von Weizsäcker has just filed a patent for a 'process to generate energy and neutrons by an explosion—e.g. a bomb', and Heisenberg is sure that the path towards the eventual development of a nuclear bomb is now open.

Although he's used to speaking openly with Bohr, Heisenberg is now sworn to absolute secrecy, so what comes out is a jumble of words and obscure allusions to reactors and bombs. Heisenberg can't judge how much Bohr already knows. Heisenberg is aware that he must be careful, as he has some powerful enemies in the Nazi system. One careless sentence could cost him everything.

Niels Bohr is not building a bomb. He has heard nothing from the American nuclear project since the previous year. Heisenberg's suggestion that the outcome of the war will be decided by the nuclear bomb is a shock to him. Heisenberg hands Bohr a piece of paper on which he's sketched out a design for a reactor.

Is Heisenberg trying to threaten him? Or warn him? Do his insinuations contain a concealed offer to work together to achieve an agreement among all physicists on both sides of the Atlantic not to build a nuclear bomb, for the sake of peace? Perhaps he is being so vague because he fears for his life. And perhaps that's why Bohr misunderstands him and is shocked to hear Heisenberg's confidence that such a bomb can be built. Why on Earth would Heisenberg ask him if he believes it is right to be researching the uranium problem during times of war, when that is precisely what he is already doing?

Their conversation does not go well. There is now too much separating the two men—too much mistrust, too many misunderstandings. After a short while, they return to Bohr's home. Heisenberg is dismayed. Bohr is highly agitated. It's midnight when Bohr accompanies Heisenberg to the tram that will take him back to his hotel.

Bohr does not attend the lecture given by Heisenberg a few days later at the German Cultural Institute—a Nazi propaganda centre. Heisenberg visits the Bohrs one more time, on the eve of his departure. They avoid any difficult questions, trying to maintain their friendship.

Heisenberg plays Mozart's Sonata No. 11 on the grand piano. The jollity of the *alla turca* movement fades quickly and it's time for them to say their goodbyes.

Bohr drafts several letters to Heisenberg after that, but never sends them. Following his return from Copenhagen, Heisenberg writes to a friend, 'Perhaps we humans will recognise someday that we really have the power to destroy the Earth completely so that we can easily create a "Judgment Day" or something close to it, by our own fault.' By the time Werner Heisenberg next visits Copenhagen, in January 1944, to check on Niels Bohr's institute, Bohr himself will be in America, working on the construction of the nuclear bomb. After the war, the two exchange polite birthday greetings, but nothing more. They never clear the air between them.

BERLIN, 1942

No bomb for Hitler

Berlin, spring 1942. Germany is no longer quite so confident of winning the war. The blitzkrieg against Russia has ended in failure. As a result of the Japanese attack on Pearl Harbor, Germany now also finds itself at war with the US. This is now truly a world war. Resources are growing scarce in Germany. Adolf Hitler orders that the country's entire industry be dedicated to the war effort. As an 'offshoot of Jewish pseudoscience', uranium research is of little interest to Hitler. His hopes for victory lie in the rockets that engineer Wernher von Braun has been firing to the edge of space from the Army Research Centre in Peenemünde on the Baltic island of Usedom.

Germany's uranium researchers need raw materials — uranium ore from Bohemia and Belgian Congo, heavy water from Norway, steel from the Ruhr Valley, and aluminium from Lusatia. They fear falling behind due to demand from the arms industry. The head of the research department at the army's high command, Erich Schumann, informs the researchers that, 'given the present personnel and raw materials situation, the nuclear power project requires resources that can only be justified if there is certainty that an application will be found in the near future'.

Schumann convenes a conference at which the researchers must explain their progress and the outlook for the success of their project to the Nazi bigwigs. Heisenberg assures them that 'the purification of U-235 will lead to an explosive of unimaginable force'.

Field Marshal Erhard Milch asks how big the bomb would have to be to destroy a city like London. Cupping his hands, Heisenberg answers, 'About as big as a pineapple'. Armaments minister Albert Speer wants to know when they might expect the nuclear bomb to be ready. Immediately, in principle, Heisenberg tells him, but in practice it will take a few more years; at least two, perhaps three or four. The war strategists are not prepared to wait that long. Speer decides that 'the nuclear bomb will not be significant for the foreseeable duration of the war'. Schumann ridicules what he calls the physicists' *Atomkackerei*, their nuclear farting about.

From that point on, uranium research is designated as a civilian project in Germany. Heisenberg and his colleagues are allowed to continue tinkering with their uranium burners, and Albert Speer even grants them a couple of million marks in funding, but he has no further interest in the nuclear bomb. Werner Heisenberg becomes the director of the Kaiser Wilhelm Institute for Physics in Berlin, making him the top scientist in the country's nuclear program—Heisenberg, of all people, who is a pure theoretician and has always avoided laboratories. He thanks Heinrich Himmler for the 'restoration' of his 'honour', and no longer dedicates his intellect exclusively to uranium research. He develops a theory of the interactions between elementary particles, which, like his interpretation of quantum mechanics, uses only observable values. It becomes known as *S*-matrix theory.

Whatever Heisenberg is up to, he's playing a dangerous game by using bloodstained Nazi funds for his research. His former student Rudolf Peierls later quotes Shakespeare, 'he must have a long spoon that must eat with the devil,' adding, 'Perhaps Heisenberg realised that no spoon was long enough.'

Heisenberg's uranium machine is located in an outhouse of the Kaiser Wilhelm Institute. Signs reading 'caution viruses' deter any unwelcome visitors. The apparatus consists of plates of uranium immersed in heavy water. The scientists bombard this with neutrons, and measure whether more neutrons are emitted than were injected. They succeed in producing energy, but never enough to raise the reaction above the critical threshold to produce a self-sustaining chain reaction. The scientists handle the radioactive substances without protection. Once, a technician's hand is burned by a flash-flame as he is pouring uranium powder into the reactor vessel. Another time, the reactor explodes; Heisenberg barely manages to escape from the room in time, and the fire brigade has to be called in. Heisenberg fiddles with the configuration of the apparatus, increasing the flow of neutrons, but still not enough to cross the critical threshold. This continues to elude German uranium researchers, including physicist Kurt Diebner, a rival of Heisenberg's. Heisenberg had ousted Diebner from the Kaiser Wilhelm Institute. Now he's working on his own reactor at the Chemical, Physical and Nuclear Test Site in the village of Gottow to the south of Berlin. The two men cannot stand each other. They prefer to compete rather than cooperate, and they squabble over the scarce uranium and heavy water resources. Diebner is the smarter experimenter of the two. He has the idea of dividing the uranium into cubes to increase the area of contact with the heavy water, enabling him to increase the production of neutrons by a couple of percentage points. Heisenberg reluctantly also adopts the cube idea.

What Heisenberg cannot know is that Enrico Fermi manages to produce a controlled chain reaction in his Chicago laboratory on 2 December 1942. The world's first nuclear reactor is up and running. The construction of Chicago Pile 1 is as crude as that of Heisenberg's uranium machine—an unprotected pile of uranium plates in a basement beneath the spectator stands of a football stadium. A bucket of cadmium nitrate solution stands ready to be thrown over the reactor in the event of an emergency. Fermi uses pure graphite, rather than scarce and expensive heavy water, to brake the neutrons' paths. This is a trick that slipped through Heisenberg's net. He tested graphite as a moderator substance, but rejected it.

The Rome-born scientist Enrico Fermi studied theoretical physics under Max Born and Paul Ehrenfest before switching to experimental physics. He's one of the few people who defy the law of the conservation of genius, as he is both a brilliant theoretician and a brilliant experimental physicist. He understands the theory behind exploding atomic nuclei, and he plans to make them useful to the war effort. His proposes poisoning German food with the fission product strontium-90. A nuclear bomb is still far in the future.

STOCKHOLM, 1943

Flight

For three years now, Adolf Hitler has reined in his violent tendencies and spared Denmark his anti-Semitic and racist laws. He upholds the fiction that he has no plans to Nazify Denmark. But after its defeat at Stalingrad and the start of the Allied blanket-bombing of German cities, the Nazi regime loses any restraint it may have had when it comes to achieving the 'final victory'. Propaganda minister Joseph Goebbels screams his slogans about 'total war': 'Now, people rise up, and let the storm break loose!'

In October 1943, during the Jewish New Year festival of Rosh Hashanah, the SS swarms onto the streets of Copenhagen. They have orders to arrest all Jews. However, there are barely any Jewish people left in the city. Georg Duckwitz, a German diplomat in Copenhagen, had warned Danish Jews of the plans to deport them to the concentration camps, and most had gone into hiding as a result. Niels Bohr and his family cross the Øresund in a fishing boat to reach neutral Sweden shortly before the SS come knocking on the door of his institute, break in, and steal the silverware. On arriving in Stockholm, Bohr appeals for help from King Gustav V for his persecuted compatriots. That same evening, Swedish radio announces that the country's borders will be open to Jewish refugees from Denmark. They take anything with wheels, from ambulances to dustcarts, to reach to the coast, hiding in churches and hospitals on the way. They cross the Øresund and the Kattegat in hundreds of fishing boats, canoes, and rowing boats. More than 7,700 Danish Jews escape the Nazis in this way.

Stockholm is teeming with Nazi agents, and is no safe place for Bohr—at least, not safe enough in the eyes of the Allies. The British physicist Frederick Lindemann, chief scientific adviser to prime minister Winston Churchill, arranges a flight to Scotland for Bohr. Before leaving, Bohr gives orders for the gold Nobel Prize medals kept at his institute in Copenhagen to be dissolved in 'aqua regia', a mixture of nitric and hydrochloric acid, so they don't fall into the hands of the Germans. After the end of the war, the gold will be precipitated out of the solution and sent to Sweden so the medals can be recast.

Niels Bohr leaves on a BOAC-operated Mosquito bomber plane, which is able to fly fast and high enough to avoid German anti-aircraft fire. Bohr has to sit in the plane's bomb bay with a mask to provide him with oxygen, and wearing a helmet with headphones to allow him to communicate with the pilot. However, the helmet is too small for Bohr's impressively sized head, and he misses the pilot's instruction to put on his oxygen mask, losing consciousness as a result. Realising something is amiss, the pilot reduces altitude over the North Sea, thereby saving Bohr's life. After being interrogated in England, Bohr is placed on a much more comfortable aircraft and flown to the US, complete with false papers issued in the name of Nicholas Baker. After reaching the laboratories of the Manhattan Project at Los Alamos in the deserts of New Mexico, he's shown round by Edward Teller. Bohr is impressed: 'You see, I told you it couldn't be done without turning the entire country into a factory. You have done just that.' The project is fuelled by fear that the Germans might be faster, and it already employs 125,000 people. One of those employees is now Niels Bohr.

Bohr has brought with him the sketched plans for a reactor that Heisenberg had handed to him in Copenhagen two years earlier. They are worthless in terms of weapons technology, according to Robert Oppenheimer, the Manhattan Project's scientific director. He also points out that Heisenberg had not travelled to Copenhagen two years ago to reveal what he knew, but to hear whether Bohr knew anything he didn't.

In Berlin, Heisenberg has now become a member of the 'Wednesday Society', a select group of intellectuals that describes itself as a 'free society for academic entertainment'. Its maximum of sixteen members includes scientists, philosophers, writers, lawyers, diplomats, and doctors — all of them men. They meet at one of their homes every second Wednesday evening. The host delivers a lecture on his area of expertise, which is followed by a discussion. In June 1943, Heisenberg speaks at the society on 'the change in the concept of reality in the exact sciences and the conclusions that can be drawn from that change'. People used to describe the world as if they were not part of it themselves, says Heisenberg. In the modern era, the era of quantum mechanics, the world can only be understood as a world that is observed, and therefore influenced, by humans. Humans alternate between those two worlds, Heisenberg continues. They cannot help secretly thinking about themselves: they are doomed to change the world they live in.

The Bomber Command of the British air force launches the 'Battle of Berlin' in November 1943. Allied planes are soon mounting air raids on the capital of the Reich around the clock, with Royal Air Force Lancaster bombers attacking at night, and US Air Force Flying Fortresses bombing during the day. The Kaiser Wilhelm Memorial Church, a symbol of German national pride, goes up in flames during the bombings on the night of 23 November 1943. Its roof collapses, and the top of the steeple breaks off. The Plancks are bombed out of their home in Grunewald. Max and Marga Planck move to the countryside to stay with friends. Writing in his diary in Paris, army captain Ernst Jünger notes that the Germans have lost the right to complain.

On 12 June 1944, a bright summer's day, Werner Heisenberg delivers another talk to the Wednesday Society, this time with the title 'On the stars'. Once again, it's an encrypted talk about his own research — he speaks about the 'nuclear fire' burning inside stars, rather than the nuclear fire he's trying to kindle just a few doors away at the Kaiser Wilhelm Institute. Earlier in the afternoon he had picked raspberries in the institute's garden to give to his guests. 'The mood was depressed,' remembers one participant. This is the 1055th meeting of the Wednesday Society, and it will be its penultimate one. On 20

July 1944, the aristocratic Wehrmacht officer Claus von Stauffenberg attempts to kill Hitler with a bomb. Four members of the Wednesday Society are arrested and executed for their involvement in the assassination plot. When he's arrested at army headquarters in Berlin, the retired general chief of staff, Ludwig Beck, who had attended Heisenberg's talk 'On the stars' only eight days earlier, asks to keep his pistol 'for private use'. 'Very well, do so, but do it immediately!' replies General Friedrich Fromm. Fromm had sympathised with the conspirators before the assassination attempt, but now tries to save his own skin by changing sides — to no avail. He's convicted of 'cowardice in the face of the enemy' before a special court, and executed by firing squad. The Wednesday Society is subsequently disbanded by the Gestapo.

As the Red Army advances on Berlin, the uranium project is removed from the bombed-out capital and distributed across various safer locations. Kurt Diebner's lab is relocated to the small Thuringian town of Stadtilm. An ultracentrifuge for enriching fissile uranium is built in Freiburg. The Kaiser Wilhelm Institute for Physics, along with Werner Heisenberg and Max von Laue, decamps to Hechingen in the Swabian Jura mountains, where it is housed in an old textile factory. Otto Hahn and the Kaiser Wilhelm Institute for Chemistry move into the neighbouring town of Tailfingen. The research reactor, the core of the uranium project, consisting of 664 finger-sized uranium cubes in an aluminium tank lined with graphite and filled with heavy water, is installed in the stone cellar of the Schwanenwirt inn in the mountain village of Haigerloch, where beer barrels and potatoes had previously been stored. Heisenberg regularly cycles the fifteen kilometres between there and Hechingen. He gathers mushrooms in the forest, admires the blossoming fruit trees, gives piano recitals for the townspeople of Hechingen, and 'can forget the past and the future for days at a time'. Elisabeth Heisenberg, now the mother of six children, stays in the log cabin on the rocky banks of the Walchensee lake, and is envious of her husband's idyllic life in Swabia.

However, Heisenberg is aware of how fragile his idyll is. Germany is losing the war. In December 1944, he travels abroad for the last time before the collapse of the Nazi state. He delivers a lecture at the Federal Technical University in Zurich—not on his nuclear research, but on his *S*-matrix theory. Some of his colleagues consider the theory an aberration, just like the unified field theory sought by Einstein. Wolfgang Pauli, now teaching in Princeton, calls it an 'empty conceptual framework'.

At the post-lecture dinner, one of Werner Heisenberg's colleagues, Gregor Wentzel, challenges him to admit Germany's defeat. Heisenberg's reply, 'It would have been nice if we had won the war,' reaches the ears of the Gestapo. The SS launches an investigation into Heisenberg, and he escapes arrest by a hair's breadth.

He returns to his cellar laboratory in the Swabian mountains, and, while the Reich collapses around him, continues to try to make his reactor work. One of his letters to Zurich falls into the hands of the American secret service. The postmark reveals that Heisenberg is working in Hechingen.

PRINCETON, 1943

Einstein mellows

Albert Einstein has made a home for himself in Princeton, the 'island of his destiny'. He is accompanied on his walks by another 'tiny demigod on stilts'—Kurt Gödel, an Austrian logician who joined the institute in 1940. After the 'annexation' of Austria, Gödel escaped and ended up in the US via an adventurous route through Siberia and Japan. Once a week, Kurt Gödel, Wolfgang Pauli, and Bertrand Russell meet at Einstein's house to spend the afternoon discussing philosophy. Pauli has been at the Institute for Advanced Study since 1940, finding Zurich to be too close to Hitler for comfort. Russell, who does not have a position at the institute and is only a visitor, gives the occasional lecture and works as a freelance writer. The group meeting at Einstein's house at 112 Mercer Street is perhaps the most distinguished circle of elderly gentlemen of science in world history.

Einstein has also come to terms with quantum mechanics. He's still not convinced by it. And never will be. However, he's given up trying to disprove it. In a letter to Max Born—now living in Edinburgh—Einstein writes, 'You believe in a God who plays dice, I in complete law and order in a world that objectively exists, and which I, in a wildly speculative way, am trying to capture. Even the great initial success of the quantum theory does not make me believe in the fundamental dice game, although I am well aware that our younger colleagues interpret this as a consequence of senility.'

Niels Bohr occasionally pays Einstein a visit at the institute, and the two old men argue over quantum mechanics—just as they always have, and yet differently now. It's no longer the battle it once was, but

Wolfgang Pauli, Professor of Theoretical Physics at the Federal Technical University in Zurich, Switzerland, November 1945

more like a cherished old ritual, a comfort for Einstein in his solitude. He's alone in his quest to find a theory that transcends relativity and quantum mechanics. His circle of friends has shrunk down to just Gödel and a couple of others. Both his marriages ended in failure. He's estranged from one of his sons, while the other is mentally ill, and he hasn't known his daughter's whereabouts for years. When Albert Einstein dies in April 1955, the chalkboard in his office at the institute is covered with equations that go nowhere.

ENGLAND, 1945

The impact of the explosion

Allied troops cross the Rhine in March 1945, along with agents of the American intelligence mission, Alsos, which has spied on the German nuclear project since 1943. Their assignment is to catch and interrogate German uranium scientists. Time is of the essence. The Soviets and the French are also after the German physicists.

'Alsos' is ancient Greek for 'grove'. The agents have been dispatched by Brigadier General Leslie Groves, the military director of the Manhattan Project. When he hears the mission's codename, he's angry at the reference to his name. Codenames are supposed to hide identities. But Groves leaves the name unchanged, as altering it would attract even more attention.

The Alsos agents search abandoned labs, gather evidence, confiscate files, and arrest scientists.

Würzburg, the city of Heisenberg's birth, is bombed by the Royal Air Force on 16 March 1945, in the final weeks of the war. Within the space of a few minutes, Lancaster and Mosquito bombers release a firestorm of demolition and incendiary bombs on the city. The inferno burns hotter than a thousand degrees Celsius. Many people run out of their cellars and try to cross the River Main to escape the fires. The fire brigade tries to help by hosing down escape routes, but the water simply evaporates in the heat, and thousands lose their lives.

In the rock cellar in Haigerloch, Heisenberg sits by his uranium machine, doing calculations. His neutron counter is showing a higher value than ever before. He places a block of cadmium next to the reactor, ready to throw it into the tank if the chain reaction runs out of control. If he only had a little more uranium, a bit more heavy water, a bigger tank—he's so close to success. But the uranium plant run by the industrial Auergesellschaft company and the heavy water plant run by I.G. Farben have both been bombed. The last remaining German troops withdraw from the Swabian Jura in April.

One afternoon in April, Heisenberg hears the engines of approaching French tanks, and decides to flee. Before leaving, he makes sure the last remaining food supplies are stored in the cellar of the textile factory, then takes his leave of his co-workers. At three o'clock in the morning on 20 April 1945, he mounts his bicycle, as there are no other modes of transport available. He's 260 kilometres from Urfeld. Heisenberg issues himself with a travel pass. He grabs a packet of Pall Mall cigarettes, just in case. He cycles for three nights, hiding from low-flying Allied planes and marauding German soldiers during the day. He witnesses the bombing of Memmingen, and passes Weilheim as it burns. He's hungry and freezing in the cold days of spring. He encounters children in adult-sized Wehrmacht uniforms. When a soldier at a checkpoint questions his self-made travel pass and makes to arrest him, Heisenberg produces the pack of cigarettes, and the soldier lets him through. On 23 April, an hour after French troops invade, Alsos agents root out Heisenberg's research group in Hechingen. Only Werner Heisenberg himself, the mission's 'target number one', is gone without a trace. In the cellar, they discover the pitiful remains of the reactor that never worked. The uranium cubes are buried in a field somewhere, and the research notes have been welded into a canister and sunk into a cesspit. The commander of the Alsos mission, Colonel Boris Pash, has the aluminium tank blown up. So this was the terrifying uranium machine that caused the Manhattan Project to rush so much? It's almost laughable.

Heisenberg reaches Urfeld exhausted, emaciated, and tattered. After hugging Elisabeth and the children, he sets about gathering supplies of food and firewood, and sealing the windows of the house with sandbags to secure them against gunfire. There are still Waffen-

SS soldiers scattered around the area, waiting to withdraw before the advance of the 7th US Army. On 1 May, Werner and Elisabeth fetch the last bottle of wine out of their cellar and raise a toast to Hitler's death. Nazism, on which Werner Heisenberg never had a clear stance, is now finished.

Now clarity is forced on Heisenberg. Colonel Pash steps onto the veranda of Heisenberg's wooden cabin on 4 May 1945 to find his 'target number one' sitting peacefully, staring out over the lake. Pash has the reputation of being a go-getter. Even before the infantry of the 7th US Army march into Urfeld, he enters the village with two soldiers to arrest Heisenberg personally—and an entire German battalion surrenders to him.

Heisenberg invites Pash and his men inside, and introduces them to Elisabeth and their children. He asks his captor how he likes the scenery. Looking at the mountains surrounding the lake, the fresh covering of snow sparkling in the spring sunshine, he says it's the most beautiful place he's ever seen anywhere on Earth. Heisenberg is relieved. His fate has been taken out of his own hands. The German Wehrmacht surrenders two days later.

Werner Heisenberg, just reunited with his family, will not see them again for nine months. Pash transports him away in a jeep. He's taken to the Alsos mission headquarters in Heidelberg. When Heisenberg is taken to the interrogation room, he discovers an old acquaintance waiting for him, sporting a US army uniform. It's Samuel Goudsmit, the Dutch physicist who last met Heisenberg six years earlier in Michigan, during his American lecture tour. Goudsmit's Jewish father and blind mother had been taken from their house in The Hague by occupying German forces and deported to Auschwitz on a cattle train. Goudsmit had appealed to Heisenberg for help at the time, but he failed to reply for months, eventually writing a letter thanking Goudsmit for his hospitality towards German scientists, and expressing his concern for the safety of Goudsmit's parents. That was all. That same year, Goudsmit became the scientific director of the Alsos mission. He's the man who's been hunting Heisenberg.

Now Heisenberg's fate is in Goudsmit's hands. He interrogates

Heisenberg, who still thinks they're having a friendly conversation, and believes he's the more knowledgeable of the two, as was once the case. He holds out his hand to his 'dear Goudsmit' for a handshake, but Goudsmit spurns it. 'Heisenberg's attitude during our conversations was particularly defiant because he believed that his work on the uranium problem was more advanced than ours, and that was the reason we were interested in him,' Goudsmit writes of the interrogation. 'Of course, we did not correct his erroneous view.'

Heisenberg is eventually taken to Paris and interned in the decrepit chateau Le Grand Chesnay with nine other scientists from the uranium club. There, he's reunited with his old student Carl Friedrich von Weizsäcker, as well as his older colleagues Max von Laue and Walter Gerlach, his hated rival Kurt Diebner, and the taciturn Otto Hahn. The Americans give the camp the fitting nickname of 'The Dustbin'. The doors are hanging off their hinges, the wallpaper is peeling off the walls, and there is no furniture except the iron bunks that the inmates sleep on.

What to do with all this concentrated brain power gathered together there? The victorious powers are unsure. One general suggests executing the German scientists, but they're too valuable for that. They are transferred first to Belgium, and then to England, where they are interned on a farm in the country of Huntingdonshire. The internment centre, housed in a red-brick building called Farm Hall, is run by the British foreign intelligence service, MI6. Heisenberg is familiar with the area, which is only one-and-a-half hours from Cambridge by bike.

The German researchers enjoy a more comfortable life in this English backwater than do most people immediately after the end of the war. Farm Hall is equipped with sports facilities, chalkboards, a billiard table, and a radio. There is plenty to eat. 'I wonder if there are microphones installed here?' Diebner considers. 'Microphones installed?' scoffs Heisenberg, 'Oh no, they're not clever enough for that. I doubt they're familiar with real Gestapo methods. They're a little old-fashioned that way.' The other internees believe him. They speak freely about physics, politics, and recent events. But the Allies are not quite as old-fashioned as the Germans think. They have provided the inmates with a radio, and have newspapers delivered by express courier, for the precise purpose of making them talk. Microphones are hidden in the walls. MI6 hears every word that Heisenberg and the others utter, and stenographers write everything down.

The physicists debate and try to work out why they're being held for so long in this luxury prison. When they ask their captors, they are simply told they are being kept 'at Her Majesty's pleasure'—English legal jargon for an indefinite term of imprisonment.

The days pass, and the German scientists have plenty of time to kill. They assure each other that they are the world's leading nuclear physicists and that the American nuclear bomb project cannot possibly have succeeded where they had failed. German physics was superior, of course.

The physicists start planning their escape. Perhaps the press could be informed of their situation? Maybe they could make it to one of their fellow physicists in Cambridge, who are undoubtedly waiting eagerly to receive their pearls of nuclear physics wisdom. They actually believe that the 'Big Three'—Harry S. Truman, Winston Churchill, and Joseph Stalin—currently meeting in Potsdam, have nothing better to do there than to discuss the fate of the incarcerated German physicists. Some are confident that their collaboration with the Nazis will not have any repercussions for them, as they are, after all, the world's leading physicists, and physics rises above politics, does it not? They plan to move to Argentina to begin their new lives.

On the morning of 6 August 1945, the sun shines down on Hiroshima. At 8.00 am, many of the city's 250,000 inhabitants are sitting down to breakfast, reading the newspaper or travelling to work or school. A pink flash of light illuminates the sky. Eighty thousand people are killed instantly. Two minutes later, the pilot of the American plane that dropped the bomb on Hiroshima looks down from an altitude of ten kilometres: 'Where there used to be a city, with buildings and everything, all we could see from our altitude was black, smoking ruins.' Tens of thousands more die later from the after-effects of the explosion.

On the evening of 6 August 1945, the German physicists are

playing rugby on the lawn of Farm Hall. They chase the strangely shaped leather ball, stumbling and laughing. It's plain to see that this British ballgame is new and strange for them. Soon it will be time to go inside for dinner.

Shortly before six o'clock, the secret service officer in charge of overseeing the German prisoners, Major Thomas Rittner, takes Otto Hahn aside and informs him that the Americans have dropped a huge atomic bomb on a city in Japan. Rittner later noted that Hahn, who had been involved in the development of chemical warfare using chlorine gas in the First World War, was 'completely shattered by the news'. Hahn sees how perfidious the argument is that the atomic bomb will hasten the end of the war—he had used the same excuse to justify his involvement in the development of chemical weapons. 'He felt personally responsible for the death of hundreds of thousands of people, as it was his original discovery which had made the bomb possible,' Rittner writes in his report for 6 August 1945.

Rittner has to give Hahn several glasses of gin to calm him down before they can join the others in the dining room and tell them the news. There is incredulity all around the dinner table. 'I don't believe a word of it,' says Heisenberg. 'I don't believe it has anything to do with uranium.' Hahn sneers, 'If the Americans have a uranium bomb, then you are all second-raters. Poor old Heisenberg.' After dinner, they listen to the BBC news on the radio. There's no more denying it. The Americans have detonated an atom bomb. Gerlach loses control, shouting and locking himself in his room until the next day, even contemplating suicide like a defeated general.

Werner Heisenberg's pride as a scientist takes a heavy blow. He spends the next few days furiously calculating, trying to work out how the Americans have managed to succeed where he, the great Heisenberg, had not. He is forced to admit that he had never completely understood how an atomic bomb might be constructed. He only thought he understood. He hadn't even managed to work out the critical mass of uranium.

At Farm Hall, the German physicists do what they had always done in Germany: bicker, complain, needle each other. They let the

Alsos agents listening in on their microphones know precisely why the Nazis' nuclear bomb program was nothing like the Manhattan Project: it was chaos. There was no plan. Werner Heisenberg and Carl Friedrich von Weizsäcker also unwittingly place on record their plan to rewrite the history of their wartime activities. They intend to try to convince the world that they deliberately prevented Hitler from getting his hands on such a terrible weapon and concentrated instead on developing a nuclear reactor, while the unscrupulous Americans built the bomb and used it. They dress up their technological failure as moral fortitude.

'The reason we didn't do it was because all the physicists didn't want to do it, on principle,' says von Weizsäcker. 'If we had wanted Germany to win the war, we could have succeeded.'

'I don't believe that,' Hahn replies, 'but I am thankful that we didn't succeed.'

On 14 August 1945, Werner Heisenberg gives a talk in front of his fellow inmates, and tells them he's calculated that the critical radius of the bomb is between 6.2 and 13.7 centimetres. He says the surface of such a sphere would shine 2,000 times more brightly than the sun when it explodes. 'It would be interesting whether objects could be knocked down by the pressure of the visible radiation,' Heisenberg concludes. He's back to being a scientist, back to the role he feels secure in. Many years later, when writing his memoirs, he has to admit, 'I had reluctantly to accept the fact that the progress of atomic physics, in which I had participated for twenty-five long years, had now led to the death of more than a hundred thousand people.'

Epilogue

We cannot observe the world without changing it. This realisation is what led Werner Heisenberg to develop quantum mechanics, and it was his dilemma. He wanted to explore the world, not change it. Nonetheless, he did change it. He had no choice but to change the world with his mighty theory, since he lived in Nazi Germany at a time when indifference was not an option. Other physicists had a similar experience. Not even the self-declared pacifist Albert Einstein was able to avoid world events. He encouraged the development of the nuclear bomb, and later regretted it. This is the dark side of the story that led from the cracks in the skin at the tips of Marie Curie's fingers to the dropping of the atomic bomb on Hiroshima.

The bright side of the story is the array of astonishingly, unbelievably clever and inquisitive people, and the way their minds interacted with each other. Quantum mechanics was such a strange theory that it could never have been formulated by one person alone. These pioneers were forced to collaborate, compete, and forge friendships and rivalries to bring it into being. This book has been born of the letters, notes, research papers, diaries, and memoirs they wrote.

True stories never have an end. But this book has had to end somewhere. The physicists in this story continued their work after 1945. But none ever made as monumental an advance as quantum mechanics or the theory of relativity. Einstein sought a Theory of Everything. So did Heisenberg. Neither was successful. But the theories they formulated a hundred years ago still stand up today. They are in our computer chips and medical devices. The arguments these theories triggered a century ago are still going on today. The objections Einstein raised to quantum mechanics are still being raised by sceptical modern physicists.

The story is not over yet.

Selected literature

I used this literature as a source, and recommend it for further reading:

PARIS, 1903

Barbara Goldsmith: *Marie Curie. Die erste Frau der Wissenschaft.* Piper Verlag, 2019.

BERLIN, 1900

John L. Heilbron: *Max Planck. Ein Leben für die Wissenschaft 1858–1947.* S. Hirzel Verlag, 2006.

Dieter Hoffmann: *Max Planck. Die Entstehung der modernen Physik.* Verlag C.H. Beck, 2008.

BERN, 1905

Albrecht Fölsing: *Albert Einstein. Eine Biographie.* Suhrkamp Verlag, 1993.

Jürgen Neffe: *Einstein. Eine Biographie.* Rowohlt Verlag, 2005.

Paul Arthur Schilpp (Ed.): *Albert Einstein: philosopher-scientist.* Open Court, 1949.

CAMBRIDGE, 1911

Finn Aaserud and John L. Heilbron: *Love, Literature, and the Quantum Atom: Niels Bohr's 1913 trilogy revisited.* Oxford University Press, 2013.

Jim Ottaviani and Leland Purvis: *Suspended in Language: Niels Bohr's life, discoveries, and the century he shaped.* G.T. Labs, 2009.

MUNICH, 1913

Florian Ilies: *1913. Der Sommer des Jahrhunderts.* S. Fischer Verlag, 2012.

MUNICH, 1914

Abraham Pais: *Niels Bohr's Times, in Physics, Philosophy, and Polity.* Clarendon Press, 1991.

BERLIN, 1920

Albert Einstein, Max Born: *Briefwechsel 1916–1955.* Nymphenburger Verlagshandlung, 1969.

Manjit Kumar: *Quantum: Einstein, Bohr, and the great debate about the nature of reality.* Icon Books, 2008.

COPENHAGEN, 1924

David Lindley: *Uncertainty: Einstein, Heisenberg, Bohr, and the struggle for the soul of science.* Anchor Books, 2008.

HELIGOLAND, 1925

Wolfgang Pauli: *Wissenschaftlicher Briefwechsel mit Bohr, Einstein, Heisenberg u.a. Band I: 1919–1929.* Springer-Verlag, 1979.

AROSA, 1925

Walter J. Moore: *Erwin Schrödinger. Eine Biographie.* Theiss Verlag, 2015.

BERLIN, 1926 (1)

Werner Heisenberg: *Der Teil und das Ganze. Gespräche im Um-kreis der Atomphysik.* Piper Verlag, 1969.

BERLIN, 1926 (2)

Paul Halpern: *Einstein's Dice & Schrödinger's Cat: how two great minds battled quantum randomness to create a unified theory of physics.* Basic Books, 2015.

GÖTTINGEN, 1926

Nancy T. Greenspan: *Max Born. Baumeister der Quantenwelt.* Spektrum Akademischer Verlag, 2006.

COMO, 1927

Louisa Gilder: *The Age of Entanglement: when quantum physics was reborn.* Alfred A. Knopf, 2008.

Carsten Held: *Die Bohr-Einstein-Debatte. Quantenmechanik und physikalische Wirklichkeit.* Schöningh, 1998.

Erhard Scheibe: *Die Philosophie der Physiker.* Verlag C.H. Beck, 2006.

BERLIN, 1930

Julia Boyd: *Travellers in the Third Reich: the rise of fascism through the eyes of everyday people.* Elliot and Thompson, 2017.

BRUSSELS, 1930

Sigmund Freud, Arnold Zweig: *Briefwechsel.* S. Fischer Verlag, 1968.

Abraham Pais: *'Raffiniert ist der Herrgott …' Albert Einstein. Eine wissenschaftliche Biographie.* Spektrum Akademischer Verlag, 2000.

ZURICH, 1931

Arthur I. Miller: *137. C.G. Jung, Wolfgang Pauli und die Suchenach der kosmischen Zahl.* Deutsche Verlags-Anstalt, 2009.

COPENHAGEN, 1932

Gino Segrè: *Faust in Copenhagen: the struggle for the soul of physics and the birth of the nuclear age.* Jonathan Cape, 2007.

OXFORD, 1935

John Gribbin: *Erwin Schrödinger and the Quantum Revolution*. Bantam Press, 2012.

PRINCETON, 1935

Adam Becker: *What is Real?: the unfinished quest for the meaning of quantum physics*. Basic Books, 2018.

Rosine De Dijn: *Albert Einstein & Elisabeth von Berlin. EineFreundschaft in bewegter Zeit*. Verlag Friedrich Pustet, 2016.

Arthur Fine: *The Shaky Game: Einstein, realism and the quantum theory*. The University of Chicago Press, 1996.

GARMISCH, 1936

Ernst Peter Fischer: *Werner Heisenberg: ein Wanderer zwischenzwei Welten*. Springer Verlag, 2015.

Werner Heisenberg, Elisabeth Heisenberg: *'Meine liebe Li!' Der Briefwechsel 1937–1946*. Residenz Verlag, 2011.

MOSCOW, 1937

Graham Farmelo: *Der seltsamste Mensch. Das verborgene Leben des Quantengenies Paul Dirac*. Springer Verlag, 2016.

BERLIN, 1938

David Rennert, Tanja Traxler: *Lise Meitner. Pionierin des Atom-zeitalters*. Residenz Verlag, 2018.

THE ATLANTIC, 1939

Richard von Schirach: *Die Nacht der Physiker. Heisenberg, Hahn, Weizsäcker und die deutsche Bombe*. Rowohlt Taschenbuch Verlag, 2014.

COPENHAGEN, 1941

David C. Cassidy: *Beyond Uncertainty: Heisenberg, quantum physics, and the bomb*. Bellevue Literary Press, 2009.

Ernst Peter Fischer: *Niels Bohr. Physiker und Philosoph des Atom-zeitalters*. Siedler Verlag, 2012.

STOCKHOLM, 1943

Joachim Fest: *Staatsstreich. Der lange Weg zum 20. Juli*. Siedler Verlag, 1994.

PRINCETON, 1943

Jim Holt: *When Einstein Walked with Gödel: excursions to the edge of thought*. Farrar, Straus and Giroux, 2018.

ENGLAND, 1945

Martijn van Calmthout: *Sam Goudsmit and the Hunt for Hitler's Atom Bomb*. Prometheus Books, 2018.

Photo credits

Tobias Hürter and Scribe would like to thank akg-Images (Berlin) for their support in selecting the images and for granting their permission to reproduce the following images:

p. 4: Marie Curie (AKG5440550; © akg-images / Science Source)
p. 27: Albert Einstein (AKG81628; © akg-images)
p. 103: Niels Bohr (AKG1048109; © akg-images)
p. 120: Werner Heisenberg (AKG5431035; © akg-images / Science Source)
pp. 206–07: Solvay Conference (AKG5468844; © akg-images)
p. 276: Erwin Schrödinger (AKG1051767; © akg-images / Imagno)
p. 325: Wolfgang Pauli (AKG6363711; © akg-images / Keystone)

Index of names and places